Space Debris Research

Jingrui Zhang · Lincheng Li · Keying Yang ·
Han Cai

# Space Debris Research

北京理工大学出版社
BEIJING INSTITUTE OF TECHNOLOGY PRESS

Springer

Jingrui Zhang
School of Aerospace Engineering
Beijing Institute of Technology
Beijing, China

Lincheng Li
School of Astronautics
Beihang University
Beijing, China

Keying Yang
School of Aerospace Engineering
Beijing Institute of Technology
Beijing, China

Han Cai
School of Aerospace Engineering
Beijing Institute of Technology
Beijing, China

ISBN 978-981-96-9569-0 ISBN 978-981-96-9570-6 (eBook)
https://doi.org/10.1007/978-981-96-9570-6

Jointly published with Beijing Institute of Technology Press
The print edition is not for sale in China (Mainland). Customers from China (Mainland) please order the print book from: Beijing Institute of Technology Press.



This Springer imprint is published by the registered company Springer Nature Singapore Pte Ltd.
The registered company address is: 152 Beach Road, #21-01/04 Gateway East, Singapore 189721, Singapore

# Contents

# Chapter 1
# Introduction

## 1.1 Overview of Space Debris

### *1.1.1 About Space Debris*

In a broad sense, the space debris includes man-made space debris and natural micrometeoroids. The space debris discussed at present usually refers to man-made space debris, which is distributed in the space-flight trajectories and loses its function. It is mainly composed of fragments and particulate matter, and is also called "space junk" because it may pose a threat to the spacecrafts in orbit or about to be launched. As defined in the *Space Debris Mitigation Guidelines* of Inter-Agency Space Debris Coordination Committee (IADC), space debris may be non-functional man-made objects, including their fragments and parts, in Earth orbit or re-entering the atmosphere [1]. According to the *Technical Report on Space Debris*[1] published by the United Nations in 1990, the full definition of space debris is: "Space debris can be all man-made objects, including their fragments and parts, whether their owners can be identified or not, in Earth orbit or re-entering the dense layers of the atmosphere that are non-functional with no reasonable expectation of their being able to assume or resume their intended functions or any other functions for which they are or can be authorized." Space debris pieces are mainly the wastes left in space after the completion of space missions that cannot be recovered, including abandoned and defunct spacecrafts, final-stage rocket bodies, as well as abandoned fuel tanks, fairings and separating units. In addition, the spacecraft fragments generated by collision or explosion in space also make a great "contribution" to the space debris environment [2].

Due to the huge benefits and driving role of space to social economy, national security, high-tech industry and other aspects, more and more countries have embarked on or kept developing their spaceflight undertakings. At the same time, the impact

[1] United Nations Publications. Technical Report on Space Debris. New York, 1999.

 
J. Zhang et al., *Space Debris Research*, https://doi.org/10.1007/978-981-96-9570-6_1

of human activities on the space environment has attracted more and more attention from the international community. Since the launch of the first man-made Earth satellite "Sputnik-1" in 1957, a total of more than 6700 spacecrafts have been put into orbit in more than 5000 launching activities by the end of 2016. However, only about 1800 of them are now in effective service, and the rest, including defunct spacecrafts and rocket bodies or their fragments generated by their disintegration, explosion and collision, constitute a wealth of "space junk" [3, 4].

### 1.1.2 Sources and Classification of Space Debris

With the development and wide application of aerospace science and technology, more and more spacecrafts are injected into space by human beings. Accordingly, the "space junk" is also increasing. The rapid growth of space debris is mainly attributed to the on-orbit explosion of defunct satellites and rocket remains, which account for the vast majority of the total mass of space objects, and the collision between them followed by their disintegration. For example, the on-orbit explosion of Fengyun-1C and the collision of Cosmos 2251 with Iridium 33 [5] led to the sharp increase in the number of LEO space shards. If leaving out the micrometeors and cosmic particles from natural sources, space debris can be classified into: defunct spacecrafts at the end of their life or upon failure, rocket body remains, mission-related debris, and the debris generated by collision between them or their explosion.[2] If the spacecraft is not deorbited after the completion of its mission, it will stay in orbit before falling down naturally due to atmospheric resistance and other factors after a long period of time, which will be hundreds or even tens of thousands of years for the objects in higher orbits. Among all the spacecrafts launched, only about one-third of them are now functional, and most of the rest have become space debris. Rocket body remains are the space debris formed when the last stage of a carrier rocket is separated from the payload and abandoned into a nearby orbit. The mission-related debris includes a great variety of wastes generated during the operation, deployment and control of a spacecraft, such as post-separation support fragments, explosive bolts, and solid rocket motor ejections. The other debris varies in size and may be from explosions, collisions, or space tests. The main components of such debris are aluminium, aluminium alloy, paint, steel, copper and titanium, in which aluminium, aluminium alloy and paint account for 81%. The ground-based hypervelocity impact test shows that the fragments produced for various reasons have different shapes, mainly including flake, block, plate, cylinder, box, rod and some irregular shapes. A detailed description of the main sources of space debris is shown in Table 1.1 [4].

As each country has a different spaceflight capacity, space activity frequency and spacecraft (rocket) passivation level, the amount of man-made space debris

---

[2] Liou JC. USA space debris environment, operations, and research updates [C]. 54th Session of the Scientific and Technical Subcommittee, Committee on the Peaceful Uses of Outer Space. Vienna: UNCO-POUS, 2017.

**Table 1.1** Main sources of space debris[4]

| Main classification | Sub-classification | Debris source |
|---|---|---|
| Mission-related debris (parts released during normal operation/ on-orbit routine operation) | Design-to-release objects (objects that have to be released) | Operated parts (such as fasteners, lens covers, cables, etc.) |
| | | Objects released by the test (e.g. pins, small balls, etc.) |
| | | A tether that was cut after the test |
| | | Other objects (released before recovery) |
| | Accidentally released objects | Aging-induced debris (such as flakes of paint and coating) |
| | | A tether system cut by debris or a micrometeoroid |
| | | An object released for safety before recovery |
| | | Liquid droplets (e.g. leaking from a nuclear power system) |
| | | Particles ejected by solid rocket motors |
| On-orbit disintegration | Intentional damage | A damage caused by scientific or military experiments (including self-destruction, intentional collision, etc.) |
| | | A before-reentry damage for minimizing the ground casualties |
| | | A damage for the safety of spaceborne equipment and stored data |
| | Accidental disintegration | An explosion caused by a failure during the mission operation |
| | | An explosion caused by instructed self-destruct system, residual propellant or battery after the mission termination |
| | On-orbit collision | A collision with a cataloged object |
| | | A collision with an uncatalogued object |
| Spaceflight systems left after mission termination | | The spaceflight systems that still stay in Geostationary Earth Orbit (GEO), Geostationary Transfer Orbit (GTO), Low Earth Orbit (LEO), Medium Earth Orbit (MEO) and High Elliptical Orbit (HEO) after the completion of their missions |

produced by different countries varies greatly. According to the data of the US Space Surveillance Network (SSN), the distribution of catalogable objects among major space powers or organizations is summarized, as shown in Table 1.2. According to calculation, the space debris produced by Russian spacecrafts (rockets) accounts for 34% of the total amount of space debris, followed by the United States (33%) and other countries (totaling 33%).

In space, there are many types of debris from a variety of sources. To better understand the hazards of space debris and make corresponding countermeasures,

**Table 1.2** Distribution of on-orbit catalogable objects among countries

| Country/organization | Spacecrafts (including those defunct) | Rocket remains and fragments | Total |
|---|---|---|---|
| United States | 1667 | 4734 | 6401 |
| Soviet Union/Russia | 1519 | 5071 | 6590 |
| China | 322 | 3665 | 3987 |
| Japan | 173 | 108 | 281 |
| India | 89 | 117 | 206 |
| ESA | 87 | 57 | 144 |
| France | 64 | 491 | 555 |
| Others | 895 | 114 | 1009 |
| Total | 4816 | 14,357 | 19,173 |

space debris is usually classified by size (diameter) according to the degree that it can be observed and identified [1].

(1) Large debris. The debris larger than 10 cm is called large debris, including defunct spacecrafts, rocket body remains and various fragments, most of which have been cataloged and tracked by SSN.
(2) Dangerous debris. The debris between 1 and 10 cm in size is called dangerous debris, because it cannot be tracked and observed and is often difficult to be effectively prevented, thus resulting in an unpredictable damage to manned and unmanned spacecrafts. The dangerous debris includes the small debris generated by explosions and collisions, and various mission-related debris.
(3) Small debris. The debris with a size less than 1 cm is called small debris, which is numerous and has a high probability of collision with spacecrafts. Especially for LEO spacecrafts, an environmental model is usually established for small debris on the basis of observation data and is described by flux. The small debris includes the debris from explosions and collisions, the combustion products of solid rockets, and the secondary debris clouds.

As shown in Fig. 1.1, the cumulative number of the debris pieces between 5 and 10 mm in size accounts for about 50% of the total number of debris pieces larger than 5 mm in LEO region. Due to its large amount and difficulty in observation, the debris between 5 and 10 mm has become the main threat faced by spacecrafts. Due to the limitations of detection techniques and observation conditions, only the debris with a diameter of at least 10 cm can be tracked and monitored internationally. For the debris with a diameter of less than 10 cm, only the short-range warning or protective measures can be pursued. For the debris between 1 and 10 cm in diameter (that is, dangerous debris), its spatial distribution is still not clear, and the passive protection measures against its impact are ineffective, so the tracking, observation, identification and prevention of this kind of debris is one of the key issues in ensuring the safety of spacecraft operation.

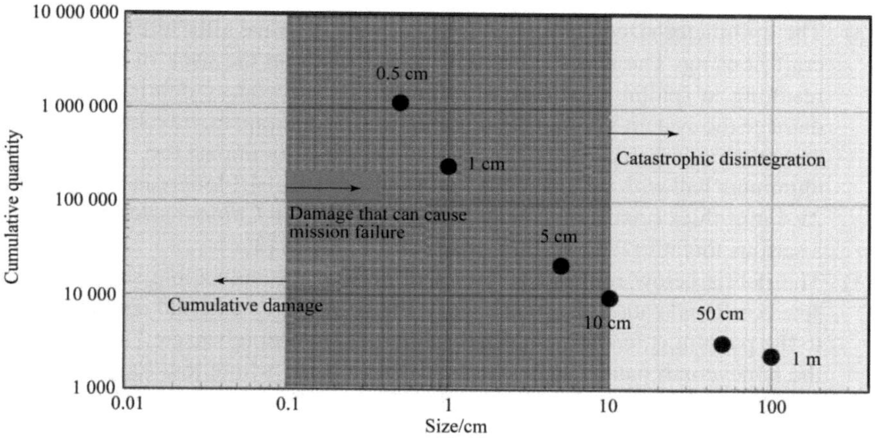


**Fig. 1.1** Cumulative quantity distribution of fragments of different sizes in LEO region (see the colored illustration) (J.C. Liou. Orbital Debris Mitigation in support of Space Situational Awareness and Space Traffic Management [R]. JSC-E-DAA-TN65688, 2019)

### *1.1.3 Hazard of Space Debris*

The collision between space debris and spacecraft at an average velocity of 10 km/s [6] is called hypervelocity collision. The shock wave formed during the collision process will cause the impact damage far larger than the size of the debris, and bring harm to the spaceflight system in many aspects. The impact of the space debris of different sizes can cause various types of damage to different parts of the spacecraft. The cumulative effect of small debris will change the performance of sensitive elements. The plasma generated by the impact will damage the power supply system of the spacecraft. The impact of larger debris will lead to the perforation, container explosion, cracking, and even structural disintegration of the spacecraft. The damage by the debris of different sizes can be divided into the following cases [7]:

(1) The debris above centimeter level is usually unpreventable. Its collision with a spacecraft will cause the destruction and disintegration of the spacecraft structure, and generate a large amount of secondary space debris. Such debris cannot be prevented, but can be accurately monitored and cataloged and can be generally avoided by evasive strategies.
(2) The centimeter-sized debris will penetrate the spacecraft housing with any possible thickness, causing damage to high-pressure RCS cylinder, propulsion system and other relatively sturdy structures, and possibly resulting in the component-level, subsystem-level and even spacecraft-level functional loss or even the explosion, disintegration and complete failure of the spacecraft. So far, such debris still cannot be accurately monitored and no effective protective measures against it are available.

(3) The millimeter-sized debris will produce a shock wave after hitting the spacecraft housing. The shock wave will be unloaded on the back of the housing, resulting in spalling and then the formation of secondary debris, which may damage the system's computer, communication equipment, traveling wave tube and other vulnerable components. The hypervelocity impact test shows that an aluminum ball with a diameter of 1 mm and a speed of 4.1 km/s can penetrate the 25.4 mm-thick honeycomb panel commonly used in Chinese satellites, posing a serious threat to the satellites' internal equipment [3].
(4) The debris below millimeter level will destroy the vulnerable surface of the spacecraft and cause direct damage to the spacecraft's external equipment and to the parts (e.g., cables, data cables, thermal control tubes, etc.) embedded in the honeycomb panel. Such small particles fill the whole low-Earth orbit and collide with the spacecrafts frequently, which can have a huge cumulative effect.

Among the above types of space debris, the debris below millimeter level is the most numerous and poses the greatest threat to spacecrafts. Specifically, its cumulative impact effect will lead to the functional degradation of photosensitive and thermosensitive materials or devices, the reduction of optical surface finish, the change of radiation characteristics of the temperature-controlled radiation surface, as well as the breakdown of atomic oxygen protection on the satellite surface. The accumulation of such impact effect by the middle and later periods of the satellite life will cause the serious chemical pollution, depression, erosion or fracture of the spaceborne optical surface, and the damage to solar array circuit, temperature control system and other vulnerable surfaces, which may bring down the satellite system function even to its failure and seriously threaten the life and reliability of the satellite. At the same time, the synergy of atomic oxygen erosion and debris impact will intensify the erosion and spalling of the damaged surface and the contamination and damage to the photosensitive surface of the satellite. The impact of small debris on satellite surface material will not only cause direct impact damage, but also result in a larger number and coverage of ejections. As a secondary contamination source, the ejections will lead to the local short circuit of solar arrays and the functional degradation of optical devices and thermosensitive materials and devices, so as to further shorten the life of satellite system and reduce its reliability [8, 9].

For the various components of the spacecraft, the hazard of the space debris impact on solar array is the most serious, followed by that on pressure vessel, thermal control material, heat pipe protection, honeycomb sandwich structure, storage battery, large parabolic antenna and other components in turn.[3] A larger spacecraft stays longer in orbit, thus facing a greater risk of being hit by space debris. As reported by the National Aeronautics and Space Administration (NASA), the satellite failures caused by debris impact accounted for 12% of the 299 on-orbit satellite failures caused by the space environment[3]. Table 1.3 shows the hazards of the space debris larger than 1 mm to the spacecraft and the countermeasures against them. Among the debris types, the millimeter-sized space debris is characterized by high impact probability

[3] KatzI. Mechanism for spacecraft charging initiated destruction of solar arrays in GEO [R]. Proc. AIAA 36th Aerosp. Sci. Meeting, Reno, NV, 1998.

and great harm, and can be dealt with passive protection. For millimeter-sized space debris, its impact and damage to the subsystems and components of the satellite are not exactly the same, as shown in Table 1.4 and Fig. 1.2.

**Table 1.3** Hazards of the space debris larger than 1mm to spacecraft and the countermeasures[3]

| Size/mm | Number/× 104 | Quantity percentage/% | Impact probability | Hazard | Countermeasure |
|---|---|---|---|---|---|
| 1–10 | 13,500 | 99.62 | Very large | Causing the component-level, subsystem-level, and even spacecraft-level functional loss or failure | The monitoring and cataloging is not applicable, but a protection structure can be installed |
| 10–100 | 50 | 0.37 | Very small | Causing the component-level, subsystem-level and spacecraft-level functional loss, and even the explosion, disintegration and complete failure of the spacecraft | Neither the monitoring and cataloging nor effective protective measures are available |
| > 100 | 1.7 | 0.01 | Very small | Causing the explosion, disintegration and complete failure of the spacecraft | The monitoring and cataloging is applicable, and active avoidance can be adopted |

**Table 1.4** Impact damage effect of millimeter-sized space debris on satellite subsystems [10]

| Subsystem level | Damage effect |
|---|---|
| Power supply | Causing the damage, perforation or disintegration of solar array and the power supply reduction or functional loss |
| Attitude and orbit control | Causing the reduction of attitude positioning accuracy or even the attitude and orbit out of control due to the bursting of high pressure vessel and the propellant leakage |
| TT&C and data management | Causing the failure of messages and commands as a result of impact-induced electromagnetic pulse generation |
| Thermal control | Causing the cracking and peeling of thermal control coating, the perforation and rupture of thermal radiator, the change of thermal and optical characteristics, and finally the functional decline or failure of thermal control system |
| Payload | Causing the performance degradation or functional loss |

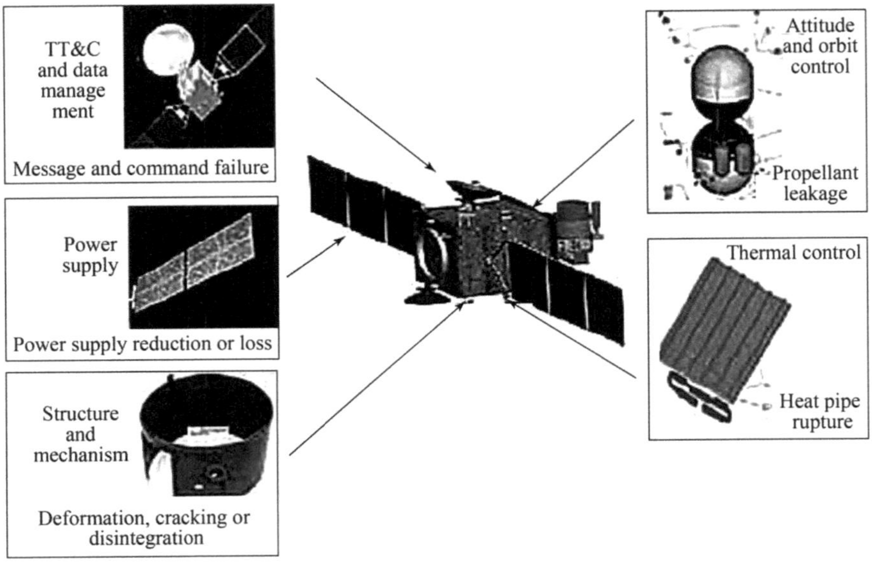


**Fig. 1.2** Satellite components vulnerable to millimeter-sized debris impact [10] (see the colored illustration)

## 1.2 Key Techniques for Space Debris Research

### *1.2.1 Detection Techniques*

In order to exploit and utilize space resources safely and sustainably, we must constantly improve the techniques for tracking and monitoring the space debris and enhance our ability to analyze and predict the space debris environment. As mentioned in Sect. 1.1, the space debris with a size of 1–10 cm poses a great threat to satellite safety. Meanwhile, such debris is difficult to be observed and identified due to the limited level of detection technology and other reasons, thus becoming the existing first problem that poses a great threat to the safety of satellite operation.

The main task of space debris detection is to detect, track, identify and confirm the space debris by various detection tools in a timely and comprehensive manner, characterize the space debris environment and monitor its changes. Meanwhile, new observation techniques are developed for warning against the approaching objects, for space object reentry prediction and for joint observation testing.

So far, there is still a large data gap for millimeter-sized debris observation. To solve this problem, the NASA Orbital Debris Program Office (ODPO) plans to further study and analyze the distribution characteristics and situational awareness of millimeter-sized space debris (especially that with a small inclination) in LEO orbit, and pay continuous attention to and detect the distribution and movement of space debris in GEO orbit.

The current detection of large-sized space debris mainly relies on ground-based radar and photoelectric telescope. The detection of medium-sized space debris mainly relies on space-based means, including space-based radar remote sensing, spacecraft surface sampling analysis and other means. However, the detection of millimeter-sized space debris (such as small debris in LEO) mainly relies on statistical models due to the difficulty in observing this debris and obtaining direct measurement data.

### *1.2.2 Early Warning Technique*

To ensure the safe operation of a spacecraft in service, the detection data can be used to predict the orbit of an object that can be tracked currently to forecast the possibility of a collision between the object and the operational spacecraft and then make a timely orbital maneuver to ensure no impact on the spacecraft. This has become a routine spacecraft safety program in foreign countries with developed space technology. Our country has also begun to consider and study the problem of large-debris collision warning by systematically analyzing and responding to the impact threat of the existing debris environment to the safe and reliable operation of the satellites to be launched or in service. The implementation of early warning technique is based on the effective establishment of a space debris environment model and the long-term prediction mainly through the calculation of collision probability.

Any space debris can be mathematically characterized. The most effective way to characterize and evaluate small debris is to establish a mathematical model due to the large amount and difficult detection of the debris. By effectively describing the distribution, movement, flux and physical characteristics of space debris, a space debris model can be used for the assessment of debris risk and hazard, the estimation of debris detection rate, the prediction of flight maneuvers for avoidance purpose, and the long-term analysis of debris removal effect. Space debris models can be divided into short-term (with a time frame of less than 10 years) and long-term (with a time frame of at least 10 years) models. Each type of model starts from the debris environment at a certain time, considers all the sources and die-outs and orbital perturbation factors, ignores the short-term effect of space object rupture, and makes evolutionary deductions. The short-term model mainly presents the current environment and is suitable for engineering design. The long-term model mainly forecasts the amount, altitude, inclination and size of space debris, which can be used as a function of time to evaluate the effectiveness of debris removal measures. Space activities and debris mitigation measures will affect the status of space debris, so the space debris models need to be constantly updated and validated.

Debris risk assessment is to assess the probability of collision between a spacecraft and space debris and its consequences, mainly considering such risk factors as space density, average relative velocity, spacecraft flight section and time. The impact effect depends mainly on the mass and composition of space debris relative to the spacecraft. To assess the risk of spacecraft penetration or failure, we need the detailed information about the spacecraft structure, including the geometry of key subsystems, their resistance to penetration, and the ability of each subsystem to withstand damage.

### *1.2.3 Protection Technique*

For small debris pieces, their orbits can't be measured one by one due to their small size and large quantity, and their impact can not be avoided by spacecraft maneuvering except by spacecraft protection. The protection design is implemented in three ways. First, the ability of a spacecraft to resist the impact of space debris is enhanced by selecting the optimal surface material, changing the structure and increasing the thickness. Second, a barrier is added to the outside of the spacecraft to reduce the impact damage of space debris to the spacecraft. Third, the vulnerable key components are placed in the positions that are not susceptible to impact. These measures are mainly applied to protect the spacecraft against the hypervelocity impact of the small debris smaller than 1 cm. For the debris larger than 1 cm in size, the increase in mass brought about by these measures will not only lead to a sharp increase in the launch cost, but also result in the extension of the mission period, the increase of system complexity and the difficulty in engineering implementation that may affect the technical feasibility. As the most important international organization engaged in space debris research, the IADC has formulated a special "Protection Manual" to guide its member states in their protection against space debris.

### *1.2.4 Mitigation Technique*

Space debris mitigation is mainly to minimize and limit the generation of space debris in the launch activities and effectively restrain the growth of space debris through the development of mitigation standards and the implementation of mitigation techniques, so as to achieve the ultimate goal of protecting the space environment and the orbital resources in outer space. The existing mitigation measures are mainly used for preventing the generation of new debris, including limiting the generation of operation-related debris and falling debris during the flight missions, taking passivation measures at the end of spacecraft life to avoid explosions as much as possible, reducing the residence time of abandoned spacecrafts by deorbiting them at their EOL (end of life), and sending the satellites that are about to fail into graveyard orbits.

Active debris removal is a method that forces the orbiting debris off the target area in a short time by means of human intervention. The techniques involved in this method can be roughly divided into four categories: push-off deorbiting, drag-increase deorbiting, autonomous deorbiting, and capture deorbiting. Push-off deorbiting is to change the orbit of debris by means of laser, ion beam, artificial atmosphere and solar radiation pressure, but may cause the generation of secondary debris due to low pushing efficiency and a limited application to orbits. Drag-increase deorbiting is to increase the flight resistance of debris by means of foam, dust, expanding ball and other devices, so that the debris can quickly reenter the atmosphere to burn and crash. This technique is mainly applicable to the Low Earth Orbit (LEO) with high atmospheric density. Autonomous deorbiting is an active deorbiting operation that relies on residual propellant or other means to deorbit the spacecraft after the mission completion. This technique may become the main way to deorbit newly launched spacecrafts at the end of their lives. Capture deorbiting is to use rigid or flexible components on the satellite platform to reliably capture the debris and control its departure from the target area. This technique is mainly applicable to large-sized space debris, especially the large defunct spacecrafts in GEO.

## 1.3 Agreements on Space Debris

The increasing of space debris has caused great harm to human spaceflight activities. The debris-related problems, such as large complexity of monitoring and early warning, high cost of removal and limited protection range, have not been solved. In addition, there are problems at the law level, such as the difficulty in defining space law and traffic liability in the *Convention on International Liability for Damage Caused by Space Objects*. As a result, the researchers are constantly seeking new regulations and policies in terms of institutional and regulatory measures to regulate space activities and related responsibilities and behaviors, while addressing the relevant problems from a technical perspective.

### *1.3.1 Institutions and Organizations*

The two important international coordinating bodies for the solving of orbital debris problem are the IADC and the United Nations Committee on the Peaceful Uses of Outer Space (COPUOS). The member states of IADC are most of the world's space powers, while the COPUOS currently has about 70 member states and discusses a wide range of space-related issues, including space debris mitigation.

1. **Inter-Agency Space Debris Coordination Committee (IADC)**

The IADC is an international governmental forum for coordinating global issues related to man-made and natural debris, seeking to mitigate and minimize the adverse impact of debris on space.

The purpose of the IADC is to: (1) facilitate the exchange of information on debris research activities among the space agencies of member states; (2) strengthen the cooperation on space debris research; (3) review the progress of cooperative activities; and (4) identify the debris mitigation options.

The members of the IADC are as follows:

- ASI (Italian Space Agency)
- CNES (French Centre National d'Etudes Spatiales)
- CNSA (China National Space Administration)
- CSA (Canadian Space Agency)
- DLR (Deutsches Zentrum für Luft- und Raumfahrt)
- ESA (European Space Agency)
- SRO (Indian Space Research Organization)
- JAXA (Japan Aerospace Exploration Agency)
- NASA (National Aeronautics and Space Administration)
- NSAU (National Space Agency of Ukraine)
- ROSCOSMOS (Russian Federal Space Agency)
- UKSpace (UK Space Agency).

The IADC consists of one planning group and four working groups. The working groups are: ① monitoring group; ② environment and database group; ③ protection group; and ④ mitigation group. The IADC is responsible for extensive international coordination and is aimed to develop the action guidelines for the countries interested in reducing the space debris formation in the future. In July 2007, the IADC adopted a set of common space debris standards focusing on four important aspects: limiting the debris release during normal operation/control, minimizing the likelihood of on-orbit disintegration, post-mission disposal, and preventing the on-orbit collisions.

2. **UN Conventions and COPUOS**

The existing UN conventions, particularly the *Convention on International Liability for Damage Caused by Space Objects*, do not distinguish between functional and non-functional spacecrafts or launch vehicles. This leads to a lack of international agreements on the naming and identification of space debris, thus reducing the willingness of countries to remove the space junk left in orbit by their spacecrafts.

In partnership with the COPUOS, the IADC has been pursuing the debris reduction measures for nearly 20 years. Based on 18 years of negotiation with the IADC, the COPUOS has agreed to adopt a set of detailed voluntary guidelines for space-sharing countries. Although quite a lot of problems remain in the field of space debris, the work of COPUOS is beginning to yield results.

3. **Spatial Data Association**

While the space powers vigorously develop their space technologies, the commercial space agencies and industrial organizations are also actively seeking ways to significantly reduce the likelihood of satellite collisions. On this basis, the Space Data Association was established with the support of the Intelsat General, the Luxembourg-based SES and the International Maritime Satellite Organization (Inmarsat). The Association initially included the major satellite communications players with assets in geosynchronous orbit, but now has expanded to not only satellite communications operators with assets in a variety of orbits, but also meteorological and remote-sensing system operators. The real-time updating of information exchange between the computers of transportation management systems enables the organizations with significant assets in orbits to be informed of the possible risk of an on-orbit satellite collision. The Analytical Graphics Inc. (AGI) provides real-time data exchange, while the Mansat in the Isle of Man is in charge of enterprise management.

The existing members of this non-profit association are:

- Spacecom Amos
- Arab Satellite Communications Organization (Arabsat)
- UK Avanti
- US Echostar
- European Eutelsat
- US Geo Eye
- US GE Satellite
- Inmarsat (Executive Member)
- International Telecommunications Satellite (Intelsat) (Executive Member)
- US National Oceanic and Atmospheric Administration (NOAA)
- Israel Parадем (the prime contractor for UK Skynet system)
- European SES Astra (Executive Member)
- US Space System Loral
- Brazilian Star One (Embratel)
- US Telesat.

The IADC, COPUOS and SDA have made great progress in improving the accuracy of space debris tracking and in expanding the capability of tiny targets tracking. To minimize the space debris, most of the space-sharing countries voluntarily follow the Guidelines. Of course, the Guidelines still need further improvement in some aspects, for example, in the establishment of national, regional or global funds for space debris removal, the creation of a new international body to manage debris minimization, the implementation of debris removal, and the revision of the *Convention on International Liability for Damage Caused by Space Objects.* According to the intention of individual countries, large debris may be removed from space. However, this may happen provided that the countries have a new attitude/posture arrangement in space. Therefore, we can only pin our hopes on the COPUOS Long Term Space Sustainability Working Group to provide some motivations, the results of which are still too early to predict.

### *1.3.2 International Agreements*

1. **Credible Removal Clause**

The International Association for the Advancement of Space Safety (IAASS) proposed a credible debris removal mechanism, namely the establishment of an intergovernmental organization by intergovernmental agreement. The organization, named International Removal Maintenance and Services for Satellites (INREMSAT), aims to establish a framework for integrated space debris development, operation and surveillance. The countries should sign legal instruments (such as treaties or international agreements) with INREMSAT to purchase the deorbiting and removal services for large space debris (including defunct satellites, upper stages and fairings of rockets, etc.) from INREMSAT. In addition, the space powers must revise their space licensing policies to include "Credible Removal" clause in their requirements for satellite launching and operation, whether by a domestic agency or a foreign agency. The clause will apply to not only satellites but also the upper stages and fairings of rockets.

The "Credible Removal" clause further requires satellite operators to demonstrate the ability of their satellite networks to autonomously and controllably re-enter the atmosphere or maneuver to the "graveyard orbits" at EOL, as well as the relevant implementation plans that have been independently audited and certified by a third party. According to the proposal, the national governments, military systems or commercial organizations should enter into service contracts with INREMSAT or similar commercial services, which, in turn, must purchase insurance for the debris removal services they provide so that the losses can be covered by the insurance company if the services fail.

2. **Space Debris Mitigation Guidelines**

In terms of debris mitigation standards, the IADC adopted the "IADC Space Debris Mitigation Guidelines" and prepared the supportive document "Technical Support for the IADC Space Debris Mitigation Guidelines" in 2002. In addition, the major space powers have formulated and introduced their own mitigation technology standards. For example, the NASA released the safety standard "Space Debris Limitation Guidance and Assessment Procedure NSS1740.14", the management guide "Space Debris Limitation Policy NPD8710.3", the supportive document "Space Debris Assessment Reference Manual" and the supportive software system "Debris Assessment Software". The ESA standard "European Space Debris Safety and Mitigation Standard", the RSA standard "Space Debris Mitigation Standard", and the JAXA standard "Space Debris Mitigation Standard NASDA-STD18" were also released. In terms of mitigation technology measures, the space powers have put forward a variety of technical measures from the aspects of space debris prevention and treatment. For example, the preventive technical measures mainly include passivation, mooring, junk orbit, reuse, and avoidance of on-orbit collision. The treatment measures mainly include deorbiting, recycling, burning and collection. The formulation and implementation of these technical and management standards and technical measures provide an important convention-based technical guarantee for effectively containing the growth of space debris and the further deterioration of space debris environment, and for protecting the environment and orbital resources in outer space.

### 1.3.3 Methods and Policies

The specific ways to solve the space debris problems include: strict pre-launching review and investigation; improving operational procedures through controlled spacecraft deorbiting, pre-deorbiting toxic gas emission, and avoiding the use of nuclear power system for eventual satellite deorbiting; developing new techniques for debris removal, actively changing the trajectory of large debris to avoid collisions, and providing more protective measures; developing new legal procedures or agreements to address the debris-related problems from the perspective of national responsibility; and establishing new procedures or mechanisms (e.g. debris removal fund) or new institutions and organizations to solve the debris removal problem.

In addition, there is still a lack of effective punitive measures to curb the continuous generation of new space debris. The existing punitive measure is merely a liability clause in principle, which means that the state whose space object damages another state's space object is deemed to be at fault and liable for another state's losses. In other words, one of the cruxes of the problem of new debris generation is the lack of truly effective incentives or penalties that would encourage or monitor countries to stop generating new debris or remove the orbiting debris.

1. **Outer Space Traffic Control**

With the improvement of the launching capabilities and the increase of commercial space activities, the space launch environment is becoming more and more complicated. The growth of commercial spaceflight interests has given birth to the concept of space traffic management. The existing commercial spaceflight portfolio includes commercial spaceline systems, cargo spacecrafts and manned flights, commercial space stations and possible hypersonic transportation systems. Commercial space activities are becoming more and more diverse and increasingly integrated with aeronautic, astronautic and outer-space transportation systems. Therefore, the problem of public safety in ground-based, space-based, stratospheric and outer-space activities must be solved systematically through outer space traffic control. The relevance here is that if we assign such responsibility to an international agency (for example, the International Civil Aviation Organization or ICAO), that agency should be able to oversee the complex aeronautic, astronautic and outer-space environments and thus manage the debris control process.

The international agency for outer space traffic control, which has been assigned a new role, should set the safety and operation standards for the missions and activities to be carried out in space and stratosphere. The agency should coordinate the development of international standards in this field, including the standards for launching sites, hypersonic transportation systems, commercial suborbital flights related to space tourism, as well as the maximum orbital altitudes of Cubesats/nanosatellites/microsatellites, and the deorbiting and falling procedures for satellites and upper stage rockets. However, it could take years for the parties to reach an agreement on the establishment of a new regulatory body, before which the parties could draft a code of conduct for international space activities and advocate better practices, safety standards and debris mitigation standards.

2. **Global Fund for Space Debris Removal**

The debris mitigation problem could be addressed by concluding national, regional and even global agreements on the establishment of an economic fund, through which a rebate mechanism could be created to compensate for the "zero-debris cleaning" launching activities and EOL cleaning activities. This mechanism provides clear incentives for space debris removal, changing the current situation that the incentives and liabilities are missing for the deorbiting or "graveyard orbit" maneuvering of debris and satellites. The creation of the fund (or the consolidation of several funds into a global fund) could provide incentives to support the development of the optimal techniques and prevent the use of a single measure, thereby favoring the debris removal activities. A 20-year "sunset clause" could be set for the fund to specify a deadline for the completion of debris removal mission. Once the mission is successful, the fund can be terminated without the need to dissolve the relevant international organizations.

The fund (or a combination of national and regional funds) could be set up in a step-by-step manner, for example, by starting with a national fund or a regional fund (for example, in Europe) and ending with a global fund in due time. Such a fund could be established by a space agency to specifically handle the space debris. By setting up a special fund, we can face the future more actively, solve the existing problems through economic measures, and avoid the lack of a financial mechanism and the passive participation of relevant countries as shown in retroactive measures.

The fund mechanism is more "economically and politically efficient" and can grant the debris removal licenses to multiple commercial organizations to facilitate the development and utilization of innovative techniques. This mechanism has the following incentives and advantages: (1) compensating for clean launching activities, and encouraging the deorbiting of upper stages and fairings of the rockets; (2) compensating for the agencies which dispose the debris at EOL; (3) establishing a "sunset clause", which defines the fund deadline and mission target; (4) promoting the competitive development of low-cost optimal techniques through the funding measures (or other incentives); and (5) requiring no dissolution of the relevant international organizations after the debris removal goal is achieved.

## References

1. Gong, Z., Xu, K., Mou, Y., et al. (2014). Space debris environment status and active removal techniques. *Spacecraft Environment Engineering, 31*(2), 129–138.
2. Ma, Z. (2017). *Evaluation of spacecraft survivability in space debris environment*. Harbin Institute of Technology.
3. Li, M., Gong, Z., & Liu, G. (2018). Advanced technology and system of space debris monitoring and removal. *Chinese Science Bulletin, 63*(25), 2570–2591.

4. Li, C., Ouyang, Z., & Du, H. (2002). Space debris and space environment. *Quaternary Research, 22*(6), 540–551.
5. Ruopu, W. (2010). *Research on space debris environment model*. PLA Information Engineering University.
6. Drolshagen, G. (2008). Impact effect from small size meteoroids and space debris. *Advances in Space Research, 41*(7), 1123–1131.
7. Qingming, Z., & Fenglei, H. (1996). Space debris environment and its hazards. *China Safety Science Journal, 6*(5), 1–2.
8. Li, K., Xie, E., Wang, L., et al. (2007). Experimental study on the continuous flashover discharge of high voltage solar arrays induced by charging. *Science in China Series E, 37*(3), 344–350.
9. Li, K., Wang, L., Qin, X., et al. (2008) Study on the charging and discharging effect of high voltage solar arrays in geosynchronous orbit. *Spacecraft Environment Engineering, 25*(2), 125–128.
10. Qi, X., & Zheng, J. (2006) Study on space debris observation technology. *Space Electronics Technology, 3*(s1), 5–8.

# Chapter 2
# Technology of Space Debris Detection

## 2.1 Introduction

In order to develop and utilize the space resources safely and sustainably, we must constantly improve the technology of tracking and monitoring the space debris and enhance our ability to analyze and predict the space debris environment [1]. As described in Chap. 1, the space debris with a size of 1–10 cm poses a great threat to the safety of satellites. In addition, such space debris is difficult to be observed and identified due to the limited detection capacity and other reasons, thus becoming the primary problem that poses a great threat to the safety of satellite operation.

The main task of space debris detection is to detect, track, identify and confirm the space debris timely and comprehensively with the aid of various detection devices, describe and monitor the space debris environment and its changes through detection, develop new observation techniques, provide the collision warning of a space object approaching the Earth, and carry out the detection tasks such as space object reentry prediction and joint observation tests [2].

The relationship between space debris detection method and debris size drawn by NASA Orbital Debris Program Office (ODPO) is shown in Fig. 2.1.[1] As can be seen from the figure, the observed debris amount increases with the orbit altitude, and a large data gap exists for the observation of millimeter-level space debris.[2]

At present, the detection of large-sized space debris mainly relies on ground-based radars and photoelectric telescopes. The detection of medium-sized space debris mainly relies on space-based means, including space-based radar remote sensing, spacecraft surface sampling analysis and other means [2]. However, the detection of millimeter-level space debris (such as small debris in LEO) mainly relies on

[1] http://ntrs.nasa.gov/search.jsp?R=20180001502.

[2] https://ntrs.nasa.gov/search.jsp?R=20160012728.

 
J. Zhang et al., *Space Debris Research*, https://doi.org/10.1007/978-981-96-9570-6_2

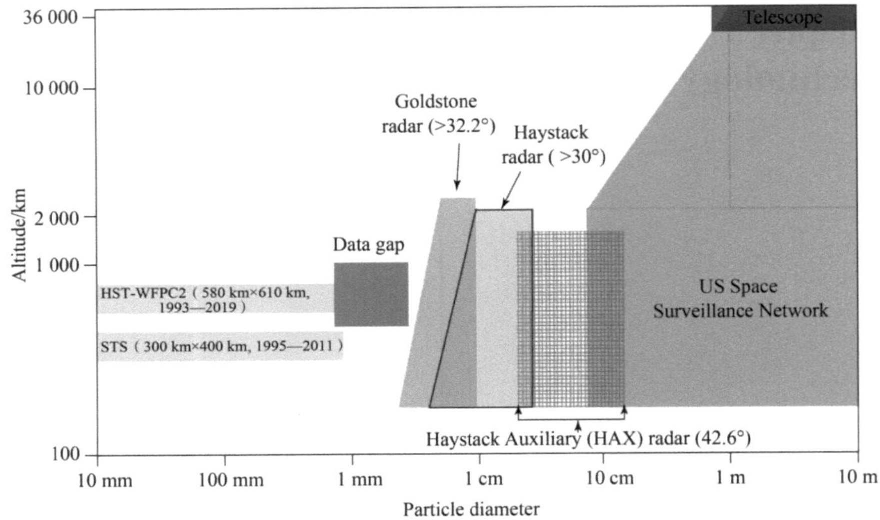


**Fig. 2.1** Relationship between space debris detection method and debris feature (the boundary in the figure is assumed) (see the colored illustration)

statistical models due to the difficulty in observing this debris and obtaining direct measurement data.[3]

In this chapter, the detection equipment, research institutions, theoretical research status and research plans for two types of detection techniques (ground-based detection and space-based detection) will be mainly presented.

## 2.2 Ground-Based Detection Techniques

Ground-based detection is to use ground-based observation equipment for the detection of space targets. The existing ground-based detection techniques are mainly classified into two categories: ground-based radar detection and ground-based optical detection. The ground-based optical detection techniques include infrared detection, photoelectric detection and laser detection. In addition to ground-based radars and ground-based optical detection means, the United States, Russia and Europe have also established their own ground-based detection networks, in which a number of ground-based devices constitute a huge surveillance system to track and observe the flying objects in space.

In terms of detection capability, the ground-based radars can detect the centimeter-level debris in LEO, and the ground-based telescopes can detect the debris as small

[3] An Overview of the Orbital Debris and Meteoroid Environments, Their Effects on Spacecraft, and What Can We Do About It? [R] Applied Space Environments Conference (ASEC), 2017.

as 10 cm in GEO. Therefore, the telescopes are more commonly used to detect the debris in GEO, while the radars are more often used to detect the space debris in LEO [3]. The ground-based detection network integrates multiple radars and various optical and photoelectric detectors, which can track and observe the objects as small as 10 cm in LEO and 1 m in GEO.

### *2.2.1 Ground-Based Radar Detection*

Ground-based radar is an important equipment for measuring the space debris. The ground-based radars used for detecting the space debris are generally mechanical scanning radars or phased array radars, as shown in Figs. 2.2 and 2.3 respectively.

The mechanical scanning radar can change the direction of antenna aiming through the mechanical control of its parabolic reflector antenna, so as to achieve the beam orientation. It is generally used for the observation of a spacecraft, upper stage of a launch vehicle, and large-sized space debris. Phased array radar is an electronic scanning radar, which uses a phased array antenna to change the phase of beam emitter by means of digital electronic technique until the radar beam points to a predetermined search area. With a fast change of beam orientation, the phased array radar can track and observe multiple targets at the same time. This radar has a strong capability of searching, tracking and observing the targets, and can effectively detect large-sized space debris in low Earth orbit (LEO).

**Fig. 2.2** Mechanical scanning radar (see the colored illustration)

**Fig. 2.3** Phased array radar (see the colored illustration)

Due to the limitation of transmitted power and working wavelength and the influence of ground noise and atmospheric loss, the ground-based radars can hardly detect a small distant space target. They can detect small space objects in LEO after the improvement of their transmitted power and working frequency, but can not effective detect the small objects in higher orbits.

Now the United States and Russia have established their own radar networks for the routine observation of large space debris. The HAX radar and Goldstone radar are the two radars used by the United States to detect small space debris. The HAX radar can detect space objects as small as 2 cm at an altitude of 1000 km above the ground, while the Goldstone radar can detect space objects as small as 0.2 cm at an altitude of 500 km above the ground and provide a statistical map of the space debris environment.

In addition, Germany's Tracking and Imaging Radar (TIRA) also has a strong detection capability. The TIRA radar has a 34 m-diameter dish antenna and operates in the L-band. The radar can detect space objects as small as 2 cm at an altitude of 1000 km above the ground. The TIRA can also be used with the 100 m-diameter receiving antenna of the nearby Effelsberg radio telescope to detect the objects as small as 1 cm at an altitude of 1000 km. The schematic diagram of this bistatic detection mode is shown in Fig. 2.4.

The European Incoherent Scatter Scientific Association (EISCAT[4]) set up a 930 MHz UHF radar and a 225 MHz VHF radar in Tromso, Norway. The association also set up a 500 MHz radar system in Longyearbyen, Svalbard, Norway, as shown in Figs. 2.5 and 2.6. The system consists of a steerable 32 m-diameter radar and a stationary 42 m-diameter radar. The main task of the EISCAT radar network is to make ionospheric measurements. By transmitting observation data back to a dedicated space debris computer, these radars can observe and count the LEO debris

[4] https://en.wikipedia.org/wiki/EISCAT.

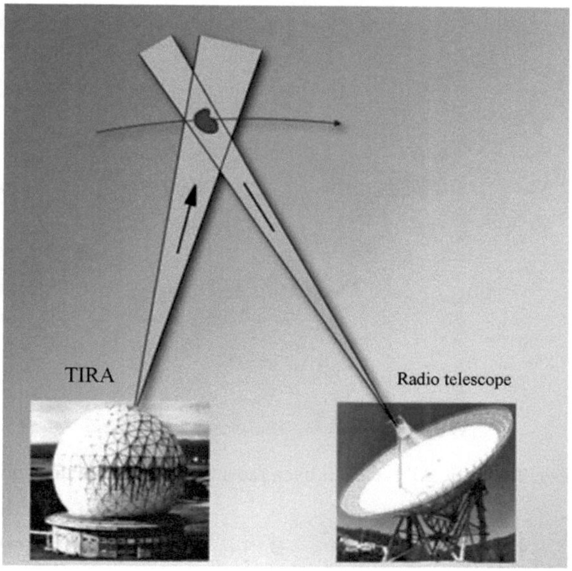


**Fig. 2.4** Diagram of bistatic radar scanning (see the colored illustration)

with centimeter-level accuracy. The test shows that the EISCAT radar system is able to continuously monitor the LEO debris clusters in a beam-parking configuration.

The Space Environment and Satellite Systems Laboratory at Stanford University, in collaboration with the EISCAT, observed the low-orbit space debris environment using a set of ground-based radar facility at EISCAT. The survey facility was used to

**Fig. 2.5** EISCAT Longyearbyen radar system (1) (see the colored illustration)

Fig. 2.6 EISCAT Longyearbyen radar system (2) (see the colored illustration)

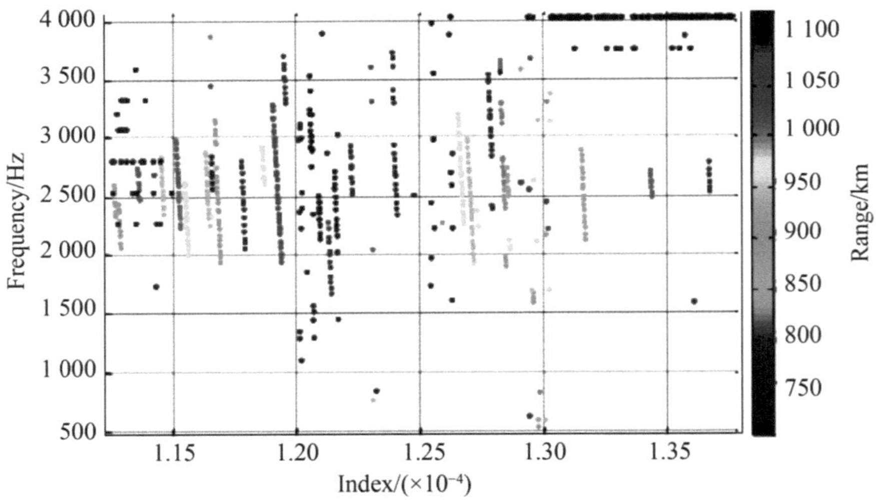


Fig. 2.7 Debris measurements from the observatory of Stanford University (see the colored illustration)

observe a debris collision event and obtained the detection results shown in Fig. 2.7,[5] in which every stripe represented a space debris. The relevant studies try to summarize the characteristics of debris change according to the detection results, and establish a model to correlate the effective cross-section of the instrumentation radar with the shape of the debris so as to predict the orbital evolution of the orbiting debris.

[5] https://sess.stanford.edu/orbitaldebris.

The Organization for Applied Scientific Research in Europe set up a 34 m-diameter dish radar in Wartberg, Germany, for the detection, tracking and imaging of space debris. The organization cooperated with many other space science research institutions such as the European Space Agency (ESA) on space debris research. This dish radar, as described in detail by ESA in its 2002 report *Detecting, Tracking and Imaging Space Debris*,[6] can use the narrow monopulse L-band for tracking and Ku band for high resolution imaging. To avoid the influence of weather conditions, the radar is housed in a radome with a diameter of 49 m. It has a peak power of 13 kW, a pulse duration of 256 ms, and a bandwidth of 800 MHz. The radar can not only detect and track space debris in orbit, but also analyze the characteristics of space debris environment and image the orbiting targets.

### 2.2.2 Ground-Based Optical Detection

The observation based on optical detection equipment (such as telescopes) is another way of ground-based space debris detection. Photoelectric telescope is an integration of telescope and photodetector and is an electron-enhanced telescope. It can collect the spectrum reflected by space objects, observe the space objects at a long distance, and image the space objects at the altitude of geostationary orbit (GSO), so as to realize the detection of large-sized space debris in MEO and HEO. The technique of optical observation has high sensitivity, but the use of optical equipment is limited by sunshine and other conditions. Clouds, fog, overcast and rainy weather, air pollution, urban background lights, and the glow of the full moon can weaken the observation capability of optical detectors or even make them impossible to observe space debris.

In the detection of space debris, optical measurement can make up for the radar's limited range, realize the measurement of space debris in MEO and HEO, and provide the high-precision data on space debris measurement. Therefore, optical measurement plays an important role in the detection of space debris and is one of the important means for determining the orbit of space debris. For example, the ESA's 1 m-diameter ground-based telescope and America's 61 cm-diameter MODEST telescope[7] are advanced ground-based photoelectric measuring equipment and are mainly used for monitoring and tracking the HEO space debris. They can detect the GSO space objects with a size of about 30 cm and a brightness of the 20th-magnitude star, and can effectively obtain the orbital distribution characteristics of space debris. However, due to the limitation of detection conditions, they can not work in all time and all weather.

As discussed at the "2017 Advanced Maui Optical and Space Surveillance Conference", the NASA's Orbital Debris Program Office (ODPO) was expected to use a meter-class autonomous telescope on Easter Island for the observation and statistics of the space debris environment in GEO, and to compare the observation results with

---

[6] D. Mehrholz, L. Leushacke, et al. Detecting, Tracking and Imaging Space Debris.

[7] https://orbitaldebris.jsc.nasa.gov/measurements/optical.html.

the simulation results obtained from a debris model, so as to accurately determine the debris orbit at different sampling frequencies. The detection telescope was expected to be able to distinguish the space debris in GEO from that in its adjacent orbits.[8]

In addition, the NASA cooperated with the ESA in 2017 to correlate the features of space debris with those of its offspring by using the observations from ground-based infrared telescopes and meter-level autonomous telescopes. They mainly adopted two methods: optical transfer measurement, and the relationship analysis based on material properties. They used three rocket bodies (Titan, BRIZ-M and Ariane) as test objects, and verified two hypotheses about the correlation between space debris features.[9]

In principle, the ground-based optical detection methods mainly include three methods: infrared detection, photoelectric detection and laser detection. The principles and characteristics of the three detection methods are presented in detail below.

#### 2.2.2.1 Ground-Based Infrared Detection

Infrared observation is a special method of optical observation using an infrared wavelength of 0.75–1000 μm. The band with $\lambda < 2.5$ μm is usually called near infrared band, the band with $\lambda > 25$ μm is called far infrared band, the band in between is called middle infrared band. The actual transmission process of infrared radiation through the Earth's atmosphere is very complicated. Due to the interaction between radiation and atmosphere, the infrared radiation is not only attenuated by absorption and scattering, but also affected by the thermal emission and refraction of the atmosphere itself and the sky noise resulting from atmospheric turbulence. Because of these atmospheric characteristics, the actual ground-based infrared astronomical observations can only be made in a few windows within the near to middle-infrared bands. The generation of infrared windows is also very complicated. Due to the absorption by water vapor molecules and carbon dioxide, the short-band infrared radiation will form several absorption bands, between which are the infrared windows. The windows between 17 and 22 μm are semi-transparent windows. The atmosphere is completely opaque to the infrared radiation windows larger than 22 μm. Only by placing a telescope on a high mountain can some infrared windows be found in this band. As the altitude increases, the density of the absorbing molecules in the atmosphere will decrease, and these windows will become more transparent and larger.

The infrared observation has a longer range than radar and can cover the targets in LEO, MEO and HEO, but has a higher cost of construction and operation than

[8] Characterizing the Survey Strategy and Initial Orbit Determination Abilities of the NASA MCAT Telescope for Geosynchronous Orbital Debris Environmental Studies [C]. Advanced Maui Optical and Space Surveillance Conference, 2017.

[9] Nasa's Ground—Based Observing Campaigns of Rocket Bodies with the Ukirt and Nasa Es—Mcat Telescopes [C]. 7th European Conference on Space Debris, 2017.

optical observation. Compared with optical band, the infrared band is wider, contains a larger wealth of information, and has greater advantages in spectral analysis and daytime observation. As a special transmitting band in the infrared band, the thermal infrared spectrum can be used to estimate the temperature of a space target. The thermal infrared imaging of a satellite can be observed both in daytime and at night. The thermal radiation will exist even after the satellite enters the Earth shadow, which may prolong the observation arc of the target. The data in the thermal infrared spectrum can also be combined with the data in the optical band to study the albedo of the target. The infrared spectrum can be used to identify the surface material of a space target. With a wider band range than visible light, the infrared spectrum can reflect more details of different surface materials. The infrared photometry, including multi-color photometry, light curve measurement and polarization measurement, can be combined with visible wavelength photometry to study the characteristics of space targets in more detail. For daytime observation, the infrared band has a lower dull background than visible light, and the near-infrared K band has obvious advantages.

In addition, compared with visible light detectors, the infrared detectors are more complicated, noisy and expensive. The ground-based infrared detection has higher requirements for optical system performance and observatory site. Specifically, the installation of infrared telescopes requires a dry environment at high altitude. In recent years, with the development of a series of infrared space-target research programs, some large special infrared telescopes and modified optical/infrared telescopes have begun to participate in the observation and research of space targets. With the progress of the research programs, more infrared telescopes will be involved in the monitoring of space targets in the coming years.

The existing representative infrared telescopes are installed at the following five observatories, which belong to the United States and the United Kingdom and represent the current state of the art.

1. **AMOS/MOTIF**

The Air Force Maui Optical Station (AMOS), located in Maui, Hawaii, is part of the Air Force Research Laboratory and is used to test and evaluate advanced optical measurement systems. The AMOS has a 3.67 m-diameter Advanced Electro-Optical System (AEOS) with advanced sensors such as adaptive optical system, radiometer, spectrometer and mid-infrared band imager that can be used to track man-made objects in deep space and collect the data about the targets.

The Maui Optical Tracking and Identification Facility (MOTIF) of North American Air Defense Command (NORAD) is also located at Maui. As the main detection facility in the Air Force's space tracking network, the MOTIF includes two 1.2 m telescopes, which are mainly used for mid-infrared and photometric data collection. With the aid of AMOS/MOTIF infrared equipment, the United States first started the infrared observation and research of space targets on Maui Island, and did a lot of preliminary exploratory work in the infrared photometry, spectral analysis and daytime observation of space targets.

2. **IRTF**

The NASA's Infrared Telescope Facility (IRTF) is a telescope with a diameter of 3 m. The IRTF is located in Mauna Kea, Hawaii at an altitude of 4200 m, and is one of the most ideal infrared stations in the world. Originally built to support the "Voyager" mission, the IRTF is now a national infrared astronomy facility in the United States, providing ongoing support for planetary science and deep space applications. Supported by a variety of auxiliary devices, the IRTF can acquire medium-high resolution spectra in the band of 0.8–25 μm. From 2006 to 2008, the NASA used the IRTF to observe and analyze the infrared spectra of space targets. In 2011, the IRTF began to participate in the daytime imaging tests of space targets.

3. **SPOT**

The Space Object Tracking (SPOT) facility, located at the Santa Cruz Proving Ground in northern California, consists of three 1 m-diameter optical/infrared telescopes that can observe space objects independently or collaboratively. In June 2012, the SPOT was put into operation. The SPOT telescope can track the LEO targets and perform photometric imaging in visible and near-infrared bands (including spectral measurement). More infrared equipment will be deployed in the future.

4. **UKIRT**

The United Kingdom Infrared Telescope (UKIRT) is the largest dedicated infrared telescope at present. It has a diameter of 3.8 m and is located in Mauna Kea. The UKIRT has a Wide Field Camera (WFCAM) with near-infrared photometric function, as well as a series of terminal equipment such as near-infrared and thermal infrared imaging spectrometers. In 2014, the NASA ODPO applied for an opportunity of using the UKIRT. Over the subsequent two years, the UKIRT spent one third of its observation time collecting the data on space targets, mainly to acquire the photometric images of known targets in geosynchronous orbit (GEO) and to detect new targets in a sky survey mode.

5. **MMT**

The Multiple Mirror Telescope (MMT), located in Hopkins, Arizona, was built in 1979. After modification, the MMT has a 0.5 m primary mirror and three replaceable secondary mirrors, providing high-resolution images in both visible and infrared bands. In January 2015, the MMT participated in a joint project between the University of Arizona and the U.S. Air Force Research Laboratory to observe the space targets in GEO.

#### 2.2.2.2 Ground-Based Laser Detection

Satellite Laser Ranging (SLR) is a satellite observation technology with the highest measurement accuracy. It is a real-time measurement technology with millimeter-level accuracy, 1–2 orders of magnitude higher than microwave radar and photoelectric telescope. The SLR can obtain the distance from a ground observation station to a satellite by accurately measuring the round-trip time for a laser pulse signal to travel from the ground observation station to the satellite.[10] In 2015, the laser ranging system of Shanghai Observatory first adopted the superconducting-nanowire single photon detector technology to successfully measure the satellite at the farthest distance of 20,00 km, providing a way to break through the laser ranging of remote small-sized space targets [4]. The space debris does not carry corner reflectors and has a low reflectivity and the laser beam incident on its surface is diffused, so the echo signal returning to the ground observation station is extremely weak. Therefore, the adoption of a large-aperture telescope, an efficient optical system, a laser system with high stability, high power and high beam quality and an echo photon detector with high sensitivity is very necessary for the laser ranging of a space target with diffuse reflection.

The correlation-distance receiver algorithm of NASA SLR2000 system divides the O-C residual plane into the grids of equal size, and compares the number of points in the grids with the preset threshold to determine whether the points are signal points. The grid size and threshold value of this algorithm are greatly affected by the observation conditions and the distance between the space debris and the station. The Mt Stromlo Station in Australia realized automatic observation data processing based on Poisson filtering. According to the Poisson filtering method, the noise points in O-C residual plane obey the Poisson distribution, while the effective signals approximate linear distribution in a short time. The effective signals are retained by judging whether the data points obey the Poisson distribution and then filtering and removing the noise points. The Graz Station in Austria used the fast echo identification algorithm to extract diffuse echo from space debris during laser ranging, but had a low success rate.[11]

The Electro Optic Systems PtyLtd (EOS) in Australia developed an active space-debris laser ranging system.[12] The system can accurately detect the debris less than 10 cm at an orbit altitude of not more than 1000 km. It has a ranging accuracy better than 1 m, an angular accuracy of 2″, a longest measurement distance of 3200 km, an orbit determination accuracy of less than 5 m, and an orbit prediction accuracy of less than 200 m (within 24 h). Currently, the system is providing routine observation services for the U.S. catalogue objects (primarily 5–10 cm debris) according to the contract requirements.

In February 2012, the German Aerospace Center (DLR) made the first detection attempt in collaboration with the GRAZ Laser Observatory in Austria, and detected the space debris 600–2500 km away with a ranging accuracy of about 0.7 m.[13]

In terms of the related research in China, the Shanghai Observatory has set up a ranging test system and conducted the ranging tests on non-cooperative targets. In July 2008, the system obtained the diffuse laser ranging data of three rocket remains.

---

[10] https://en.wikipedia.org/wiki/Satellite_laser_ranging.

[11] https://www.researchgate.net/publication/237298105_SLR2000_PROJECT_ENGINEERING_OVERVIEW_AND_STATUS.

[12] https://www.eos-aus.com/space/satellite-laser-ranging/.

[13] https://www.dlr.de/dlr/en/desktopdefault.aspx/tabid-10212/332_read-2662/#/gallery/4788.

The Yunnan Astronomical Observatory started the diffuse laser ranging research on space debris in January 2008, and received echoes from rocket remains in June 2010. Due to the low repetition frequency of the adopted ranging methods, these stations obtained only a small amount of ranging data in the laser ranging tests. Subsequently, the research on the laser ranging of space debris at high repetition frequency was started in China. The Shanghai Observatory realized the laser observation of space debris at high repetition frequency by using an all-solid-state laser with a repetition rate of 200 Hz and a power of 50 W supported by a low noise detector, a range gate generator with nanosecond-level control accuracy and a high-efficiency spectrum filter. In 2013, the Changchun Satellite Observation Station started the diffuse laser ranging research on space debris by using a high-power laser with an output power of 30 W and an emission frequency of 500 Hz, and successfully realized the detection. So far, several laser ranging stations in China have realized the detection of space debris using the laser ranging technology. However, most of the detected debris is about 1 m in size, and the ability to detect small debris still needs to be further improved [5].

### *2.2.3 Ground-Based Detection Network*

Since the 1960s, the United States and the Soviet Union, the two superpowers, have invested heavily in the establishment of a huge space surveillance system - missile early-warning and satellite target surveillance system - for the purpose of world domination and the need of their own security, while devoting their efforts to the development of their own space systems and strategic nuclear weapons. The system makes use of the most advanced information acquisition technology with the highest performance to continuously measure all the flying bodies in space, ensuring that the condition of the flying bodies can be known at any time. In the space surveillance systems of the two countries, dozens of radars and various optical and photoelectric detectors can track and observe the 10 cm objects in LEO and 1 m objects in GEO.

The United States, Russia and Europe have established their own ground-based detection networks, as described in detail bellow:

1. **The United States space surveillance network (SSN)**

Since its inception in the early 1960s, the SSN has been the earliest and largest observation system that can observe the space debris larger than 10 cm in LEO and larger than 1 m in GEO. In order to continuously track the moving trajectories of space debris, the SSN has set up 25 observation stations around the world, which can continuously track and observe the space debris at an orbit altitude of more than 600 km.

The SSN consists of the ground-based radar detection systems and ground-based photoelectric/optical detection systems distributed all over the world. Among all the SSN equipment, the radar has an active detection ability, but its range is restricted.

The photoelectric detector has a very long range but cannot meet the requirement of all-time and all-weather operation. The two detectors have their own advantages and disadvantages and complement each other to form a complete space surveillance system. The basic task of SSN is to provide information about the objects orbiting in Earth space, including the orbital parameters, size and shape of each object, as well as other data useful for determining the uses of the object.

The SSN observation radar is ground-based radio radar. The debris diameter that can be observed by radar depends on the adopted radio frequency. The higher the frequency, the smaller the observable debris diameter (or the higher the resolution).

In the early stage of SSN development, the radar frequency band adopted by SSN was UHF and could only observe the debris larger than 10 cm in diameter. In 2009, the U.S. Air Force began to plan the development of a new space debris observation system called "Space Fence" as the replacement of SSN at a cost of about $3.5 billion. Operating in the S band, the Space Fence is expected to improve the observation accuracy by half an order of magnitude, that is, to detect the debris more than 2 cm in diameter. In addition, two other radars are also used in SSN, including: Haystack radar and Goldstone radar, which can observe the debris 2 cm in diameter, but can only track local areas.[14]

2. **Russian space surveillance system**

In order to monitor and track the space environment, the Soviet Union/Russia established a space surveillance system,[15] including Space Surveillance Center, among its armed forces. The technical means and methods of Russian space exploration also mainly include ground-based radar detection systems and ground-based photoelectric/optical detection systems, which can continuously search the space, find and track various military spacecrafts, and measure the orbital parameters of satellites and their changes. "Skylight" is a typical active ground-based photoelectric space monitoring and tracking system used by the Russian space forces, also known as Nurek Space Photoelectric Observation Station. Located in the Sanglok mountain area on the Pamir Plateau at an altitude of 2200 m on the Tajikistan-Afghanistan border, the system is an indispensable auxiliary support means of Russia's strategic early warning system and can effectively fill the gap of deep space surveillance network. In addition to acquiring the position and ephemeris data of the space targets, the Russian space surveillance system also measures and catalogs the characteristic parameters of the space targets such as their size and state. By using the multi-spectral observation technology, the system is able to acquire the information about the makeup (shape and material) of a HEO satellite.

3. **European radar observation system**

The Tracking and Imaging Rader (TIRA) system[16] in Germany is now operational and can observe the regional debris larger than 2 cm in diameter. The GRAVES

---

[14] https://www.nasa.gov/sites/default/files/files/OrbitalDebrisProgramOffice.pdf.

[15] https://www.wemedia.ifeng.com/60668994/wemedia.shtml.

[16] https://www.esa.int/Our_Activities/Operations/Space_Debris/Scanning_and_observing2.

bistatic radar[17] of the French Ministry of Defence, which was put into operation in 2009, can observe the regional debris larger than 1 m in diameter.

Considering the status quo of ground-based detection described above, the current ground-based observation mainly presents the following characteristics:

The observation capability of ground equipment is restricted by many factors, such as the transmitted power and working wavelength of the radar, as well as ground noise and atmospheric loss. Under these restrictive conditions, the ground-based observation stations have a relatively high requirement for site selection and a restricted observation time, and generally can not work in all time and all weather.

Currently, the radars can detect the centimeter-level debris in LEO, and the ground-based telescopes can detect the debris as small as 10 cm in GEO. However, these observation capabilities need to be improved in order to complete the tasks such as the tracking, observation and cataloging of space debris. The bistatic observation or the combination of multiple observation means can acquire more accurate observation results, and thus will gradually become a more important means of ground-based observation.

In terms of theoretical research, the ground-based detection imaging technology pays more attention to the efficient extraction and analysis of the observation information. Advanced information processing techniques, such as statistics and image processing, can be used to acquire as much valuable information as possible from limited observation results for the statistics and estimation of space debris.

## 2.3 Space-Based Detection Techniques

Space-based detection is the detection of space debris by means of the observation equipment and detectors mounted on space-based platforms. It is an effective method to detect small space debris. Without the influence of the Earth atmosphere, the space-based detection can work at a higher signal frequency for the "close-range" and high-precision observation of space debris, especially small and medium-sized space debris that cannot be detected by ground-based systems. However, the space-based detection is limited by volume, mass and power, and its cost is high.

In terms of measurement type, the space-based methods of space debris detection can be divided into three main types: space-based remote sensing detection, space-based direct collision detection, and the sampling analysis of the recovered spacecraft surface. Among them, the space-based remote sensing detection methods are active detection methods, including optical detection, radar detection and so on. The space-based direct collision detection and the sampling analysis of the recovered spacecraft surface are passive detection methods and are the common forms of early space-based detection. In a passive detection method, the equipment is exposed to the space environment at first and recovered completely (or partially) after a period of time, and then the data about space debris can be obtained by analyzing the impact pits

[17] https://www.esa.int/esapub/bulletin/bulletin133/bul133f_klinkrad.pdf.

on the spacecraft surface. The examples of passive detection include: LDEF satellite launched by the United States, solar panels recovered by the US Hubble Telescope, porthole and radiator of the US space shuttle, EURECA platform launched by the ESA, Russia's "Salyut" space station, Japan's SFU and so on.

So far, the observation of millimeter-level to centimeter-level space debris has always been a difficult problem. Such a target is too small for ground-based observation and is beyond the sensitivity of a space-based impact detection device. The effective in-orbit remote detection may also be impractical for the space debris particles smaller than a few millimeters in diameter. However, the in-situ impact technique can effectively capture such debris and characterize the size and material of the debris as well as its orbital distribution and dynamics, except for the following problems in the use of this technique. First, the impact probe can only pick up the debris that intersects its orbit. Second, the range of the measured environment and the statistical validity of the data depend on the total exposure area and exposure time of the probe. Finally, in order to identify the specific source of the impact particles (space debris or micrometeors), the methods for measuring the velocity vector or composition of the impact particles, as well as the knowledge on their velocity vector are needed to determine their pre-impact orbital parameters.

The United States, Europe and Russia have launched some research programs on the space-based sensing of space debris. In order to accurately track smaller and more distant targets in real time, the United States developed the space-based space target monitoring systems represented by the Space-based Space Surveillance (SB-SS) system deployed in LEO and the Orbit Deep Space Imager (ODSI) system deployed in GEO. In 2010, the SB-SS system was deployed in LEO to provide the data on target location and orbital maneuver and identification to the Joint Space Operations Center (JSPOC). In 2014, two regional space imaging satellites for GSSAP were deployed in GEO. In 2016, another two regional space imaging satellites were deployed in GEO. The small-cost satellites planned to be launched in 2017 were included in the follow-up SBSS program. ORS-5, also known as SensorSat,[18] was built by the MIT's Lincoln Laboratory to scan the geosynchronous corridor in LEO. In the system, a maneuverable satellite drifts above and below the GEO to take pictures of the satellites in GEO. The two planned satellites were launched in 2014 and 2016 respectively. The ORS-5 was successfully launched aboard a Minotaur4 rocket from the SLC-46 launch station at Cape Canaveral Air Force Base at 14:04 Beijing time on August 26, 2017. In the future, the United States will continue to improve its space monitoring network. In terms of space-based observation, the Geosynchronous Space Situational Awareness Program (GSSAP) has been newly proposed. As reported on March 5, 2019, the Space Situational Awareness Technology Validation Satellite S5 developed by the U.S. Air Force was launched into geostationary orbit on February 21 as the secondary payload of the Falcon-9 rocket to demonstrate the low-cost space situational awareness constellation technology. The main sources of space-based

[18] https://news.mit.edu/2017/mit-lincoln-laboratory-sensorsat-will-monitor-activity-in-geosynchronous-belt-1129.

observation information in the United States include Space Surveillance Network (SSN), recovered spacecraft surface, Hubble Space Telescope and space shuttle [3].

In 2008, the ESA launched the space situational awareness system program[19] with an investment of 45.6 million Euros, in which 14 member states took part. The system, which consists of an electromagnetic fence and four telescopes, is able to monitor the objects at an orbit altitude of less than 1000 km and provide the services related to spatial searching and tracking data, space situational awareness, near-Earth object data and space weather. So far, more than 17,000 pieces of space debris have been tracked and catalogued by the system.

Now Russia has built the International Scientific Optical Network (ISON), which consists of 30 observation stations equipped with more than 100 telescopes and covers 14 countries and regions [6]. The ISON system can provide the following main services: tracking, observation, search and discovery of the space debris in MEO and HEO; debris cataloging and debris feature detection; warning of a collision with satellites in MEO and HEO; motion model and orbit evolution; and tracking, observation, search and discovery of the asteroids. The system's global station distribution fills America's gap in space object cataloging system. So far, more than 16,000 pieces of space debris have been tracked and catalogued by ISON [7].

In addition to ESA and other scientific research institutions, the rapidly developing foreign commercial satellite companies have also started relevant research in recent years with an increasingly strong demand for space target detection and tracking and space situational awareness.

In 2009, the Space Data Association (SDA)[20] was founded by several commercial satellite companies such as Inmarsat, Intelsat, SES and Eutelsat, proposing a new model of space target monitoring and space situational awareness based on the data and cost sharing among participants. Meanwhile, the Com-Spoc, with the support of AGI, launched a paid service for space data information. So far, the company has set up 28 telescopes around the world and acquired the orbit information of more than 5000 space objects. It can be predicted that in the future, a new model of space-based detection based on the data sharing, cost sharing and cooperation among institutions will become the mainstream with the rapid development of microsatellites and commercial spaceflight and the gradual normalization and legalization of space target impact avoidance and space debris removal. Our country has also successfully demonstrated the feasibility of developing a space-based radar system for space debris detection and engaged in the research on space-based direct detection [8].

[19] https://www.esa.int/Our_Activities/Operations/Space_Debris/Scanning_observing.

[20] https://www.space-data.org/sda/.

### 2.3.1 Space-Based Remote Sensing

Space-based remote sensing is mainly to detect space objects by using the radars or optical telescopes aboard the satellites, spacecrafts, space stations and other platforms. Since the space-based remote sensing is carried out in space, the detection process is not interfered by the atmosphere. In addition, the distance between the detection equipment and the target can be very close. Therefore, the the space-based remote sensing has a high detection resolution and can be used for the detection of small and medium-sized space debris.

According to different detection means, the space-based remote sensing methods can be divided into space-based optical detection and space-based radar detection. However, in recent years, the existing traditional detection methods have been cross-linked and fused by many scholars and even combined with some new methods into new space-based detection techniques, most of which are still in the stage of theoretical research and model testing.

### 2.3.2 Space-Based Optical Detection

Space-based optical detection is a detection means with high sensitivity and resolution that uses the optical electron telescopes aboard space-based platforms for space debris observation. However, since both the satellite platform and the observed target are moving at high speed, the observation process is limited by the platform location and the time frame available for observation and thus the observation efficiency is low. Therefore, the actual use of this detection means has certain limitations.

Figure 2.8 shows the schematic diagram of the US space-based detection system and a sketch of the SBV probe aboard MSX spacecraft that is detecting an object in GSO. As early as 1996, the United States launched the "mid-course space test satellite" MSX. The main equipment on board included a space infrared imaging telescope, ultraviolet and visible light cameras and space-based visible light sensors. The main task of MSX was to detect and track the midcourse of a missile, and give early warning of the missile midcourse. In 1997, the technical verification of the program was completed, and the integration of the program or technique into SSN was started. The MSX was officially put into operation in 1998 and retired in 2008. In its service period, the MSX demonstrated a new generation of detector techniques for missile warning and defense use, and collected and analyzed valuable background data and target data. Its mature techniques will be transferred to the next generation of space-based space surveillance systems.

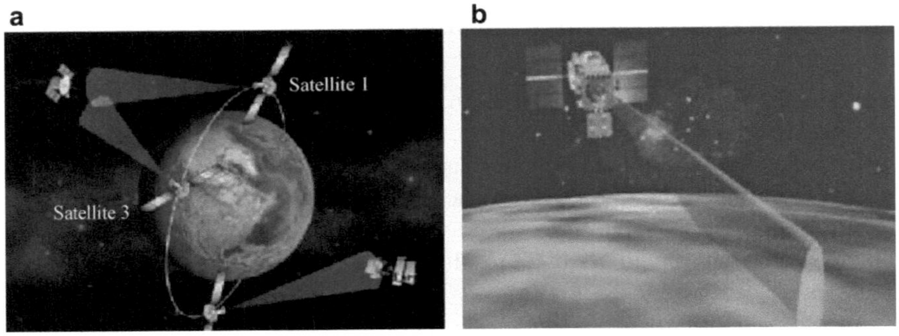


**Fig. 2.8** Schematic diagram of SBSS System (a) and a sketch of the MSX spacecraft Detecting an orbit in GSO (b) (see the colored illustrations)

### *2.3.3 Space-Based Radar Detection*

Space-based radar detection is one of the development directions of space target surveillance thanks to its complete theory, mature technology and unique advantages in space-based detection. The space-based radar systems for space debris detection reported so far include: the radars on board the US International Space Station for monitoring the orbiting targets, the microwave radars on board the minisatellite group developed by CNES, Russia's millimeter-wave phased array radars, Canada's space surveillance radars, and the phased array radars for space target tracking developed by ESA. The research on space-based detection in China mainly includes the demonstration of the feasibility of developing a space-based radar system for space debris detection [8].

Space-based microwave radar and laser radar are the focuses of current space debris detection research.

In a laser radar, the laser is used as the radiation source to extend the working band of the radar to the range of light wave band, and the laser technology is used to detect the space targets. Laser radar is an effective means of space target detection thanks to its high detection resolution and high positioning accuracy, as well as small loss (because the laser does not pass through the atmosphere). However, the laser technology is not mature at present, and is only used for the detection of short-distance space targets in space-based detection.

Space-based microwave radar is an electronic device that uses radio waves to determine the orbital parameters of space targets. Since the influence of the Earth's atmosphere is absent, this radar can detect space debris with high-frequency electromagnetic waves. Therefore, a space-based microwave radar can detect remote small-sized space debris with low transmitted power and small-sized antenna, and is especially advantageous in the detection of small and medium-sized dangerous space debris to which no effective observation means are available at present. The US space-based millimeter wave radar can be mounted on the space station or a platform close to the space station. At first, the infrared observation system provides the original position information of a 4–8 mm space target at a distance of no more than 25 km from the space station. Then, the millimeter wave radar uses various advanced techniques to track and predict the target, and provides NASA with the information about collision and early warning [9].

The radars that are aboard the International Space Station to monitor dangerous targets in orbit, as thoroughly argued in the United States, include[21]: the Ka-band millimeter wave radar developed by Texas A&M University and the Ku-band space-based radar jointly developed by Johnson Space Center and Rockefeller Engineering Science Company. The Ka-band radar is mounted on the space station or a platform close to the space station, and works in the following way. At first, the infrared detection system provides the original position information of a 4–8 mm space target at a distance of no more than 25 km from the space station. Then, the Ka-band radar uses various advanced techniques (including phased array, single pulse tracking, pulse compression, high power transmitter, low noise receiver, pulse accumulation signal processing and other techniques) to track and predict the target, and provides NASA with the information about collision and early warning. In the Ka-band radar system, two to three ground-based radar stations are relied on to provide the orbit prediction information for a target with a diameter of 2–10 cm in LEO. At first, the fixed electronic beams are used to detect the target crossing the orbital plane at the predicted position, and provide the information about the time for the target to cross the orbital plane, the target's pitch angle, the range to the target and the range rate. Then, this information is transmitted to the ground stations. By using the processing gain greater than 30 dB, the ground stations obtain the latest accurate orbit information of the space target with a diameter greater than 2 cm and a distance of 400 km, and judge the possibility of collision, and then send the maneuver avoidance information to the space station.

Vympel, a Russian company, has developed a low-power-consumption space-based radar concept for the detection of small-sized (millimeter level) space targets.[22] Working at a base frequency of 95 GHz, the radar has a phased array antenna composed of 7 antenna elements, in which the beamwidth of each scanning sector is 4 mrad and the scanning sectors are vertical to the platform orbit.

A concept of space target probe has been studied by Canada Davion Systems Ltd. The probe can be mounted on a spacecraft as its component, or operate as a separate vehicle near the spacecraft. The probe can build a database for the targets in the crossing orbits. Based on the database, the possibility of collision can be predicted. At first, a stationary beam matrix is established, in which the beams are fan-shaped and arranged like the spokes of a wheel. The pitch angle of the beams perpendicular to the wheel plane is wide, while the azimuth angle of the beams parallel to the wheel plane is narrow. Since a space target usually traverses at least two beams, the radial velocity of the target and the time interval required for traversing two beams can be

---

[21] Anzmeador P D, Vilas F. In-situ measurements of the orbital debris environment utilizing the Midcourse Space Experiment (MSX), 1994.

[22] Tolkachev A A, Zolotarev M M, Loukiaschenko V I, et al. Space Based Radar to Observe Space Debris, 1998.

deduced from the two echoes obtained after traversing the beams. Thus the position and velocity of the target traversing the beams can be determined. According to the reports on space-based radar systems for space debris observation, most of these systems work in millimeter wave band.

In fact, the millimeter wave band is particularly suitable for space-based space debris observation radars because there is no atmospheric absorption effect on millimeter waves in outer space. The millimeter band radar has a short wavelength and can observe the space debris of small size easily with high measurement accuracy. This radar has a small volume and mass, and thus fits the space-based platforms very well. However, due to the difficulty in scheduling and calibrating a space-based observation radar system and correlating the debris observation data with the cataloging database, the application of the system has not been reported.

## 2.4 Development Trend of Space-Based and Ground-Based Detection Techniques

The space debris detection techniques presented in this chapter can be summarized as the framework shown in Fig. 2.9.

Based on the existing means of space debris monitoring and imaging, the relevant Chinese institutions and scholars have put forward several suggestions for developing key technologies in the space surveillance system [3], including the following seven suggestions:

(1) Construction of ground-based detection system

A huge surveillance system composed of multiple ground-based radars and ground-based telescopes should be built to form an effective ground-based detection network. At the same time, the multi-band radar system should be upgraded and the bistatic

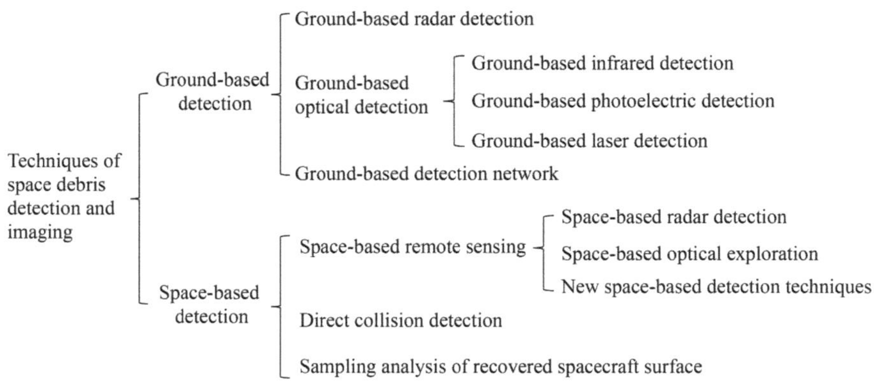


**Fig. 2.9** Techniques of space debris detection and imaging

detection mode should be further explored to effectively improve the accuracy of space debris detection.

(2) Acquisition and identification of space target characteristics

The characteristics of space targets, including photoelectron spectroscopy characteristics, motion characteristics and material characteristics, should be accurately acquired to provide basic information for debris orbit determination, space situational awareness and debris removal. The research priorities are the reflection characteristics of space targets to radar wave and visible light, the influence of debris motion state on photoelectric reflection characteristics, as well as the material composition of debris.

(3) Processing of space debris imaging information

Different image recognition algorithms and statistic means should be used to identify the debris quickly and accurately, so as to catalog and estimate the debris.

(4) Determination and prediction of space debris orbit

According to the real-time monitoring data of space debris, the debris orbit should be determined and predicted, and the influence of different measurement data and orbit determination methods on the orbit determination accuracy should be studied, so as to optimize the requirements for data sampling period and arc during the orbit determination and prediction and improve the accuracy of orbit determination and prediction and the efficiency of the system.

(5) LEO space debris engineering modeling

A breakthrough should be made in the backtracking of historical orbital disintegration events within an altitude of 2000 km and in the modeling of space debris environment in the Earth-fixed coordinate system to analyze the model errors and sensitivity factors and establish an autonomous space debris engineering model whose basic function and flux prediction accuracy are better than the latest version of ORDEM series (ORDEM3.0). Meanwhile, the effects of giant minisatellite constellations and emergencies on space debris environment should be studied to improve China's capability of autonomous spacecraft debris risk assessment and provide a basis for space debris protection engineering.

(6) HEO space debris engineering modeling

The engineering model of LEO space debris should be extended to an altitude of 36,000 km to provide a basis for GEO impact risk assessment, guide the deorbiting operation of large spacecrafts in GEO at the end of their life, improve the in-situ utilization rate of orbital resources in GEO, and reduce the probability of debris impact in GEO.

(7) Spacecraft reentry prediction and risk assessment

Based on the information on spacecraft attitude, orbit, shape and material, the aerodynamic force, aerodynamic heat and disintegration of spacecrafts during reentry should be studied to provide an accurate reentry prediction.

## References

1. Qi, X., & Zheng, J. (2006). Study on space debris observation technology. *Space Electronics Technology, 3*(s1), 5–8.
2. Wang, R. (2010). *Research on space debris environment model*. PLA Information Engineering University.
3. Li, M., Gong, Z., & Liu, G. (2018). Advanced technology and system of space debris monitoring and removal. *Chinese Science Bulletin, 63*(25), 2570–2591.
4. Deng, H., Wu, Z., Tang, K., et al. (2016). 2015 Shanghai observatory report on satellite laser ranging observation. *Annual Journal of Shanghai Observatory, Chinese Academy of Sciences, 2016*(1), 21–32.
5. Yu, H., Gao, P., Shen, M., et al. (2016). Analysis on the detection capability of space debris laser ranging. *Astronomical Research & Technology, 13*(4), 416–421.
6. Dai, K., Feng, Z., & Wan, X. (2016). Review on the development of Russian space situational awareness system. *Journal of China Academy of Electronics Science, 11*(3), 233–238.
7. Zhai, D. (2016). *Study on key technologies of diffuse laser ranging*. Graduate University of Chinese Academy of Sciences (Yunnan Observatory).
8. Yuan, Q. (2005). *Space microdebris detector*. Graduate University of Chinese Academy of Sciences.
9. Yuan, W., Zeng, L., Zhang, B., et al. (2012). Study on space debris monitoring technology. *Modern Radar, 34*(12), 16–19.

# Chapter 3
# Environmental Model of Space Debris

## 3.1 Orbital Perturbation and Orbit Prediction

### *3.1.1 Orbital Elements*

The orbiting status of a space target can be described by six orbital elements, which can uniquely determine the orbit of the target, including the position, direction and shape of the orbit, namely the position and velocity of the target in orbit. The commonly used orbital elements are classical orbital elements, which are represented by $a$, $e$, $i$, $\Omega$, $\omega$ and $\upsilon$. They constitute a group of vectors, which are called state vectors in generalized coordinates. Compared with the description method based on radius vector and velocity, this method is more intuitive. Each of the classical orbital elements is defined as follows:

(1) Semi-major axis $a$: half of the major axis of the orbit, which determines the size of the orbit. In circular and elliptical orbits, $a > 0$. In parabolic orbits, $a = +\infty$. In hyperbolic orbits, $a < 0$.
(2) Eccentricity $e$: eccentricity of the orbit ($e \geq 0$), which determines the shape of the orbit.
(3) Orbital inclination $i$: the angle between the orbital plane and the equatorial plane, which can also be defined as the angle between the Earth's rotation axis (i.e. the line between the Earth's core $O$ and the North Pole) and the outer normal line $H$ of the orbital plane. The range of the angle $i$ is $0 \leq I \leq \pi$. If $i < \frac{\pi}{2}$, the space target will travel in an easterly direction and the orbit shall be called prograde orbit. If $i > \frac{\pi}{2}$, the space target will travel in a westerly direction and the orbit shall be called retrograde orbit. If $i = \frac{\pi}{2}$, the space target will pass over the north and south poles and the orbit shall be called polar orbit. If $i = 0$ or $\pi$, the orbit shall be called equatorial orbit.



J. Zhang et al., *Space Debris Research*, https://doi.org/10.1007/978-981-96-9570-6_3

(4) Right ascension of ascending node $\Omega$: The angle of eastward rotation from the vernal equinox to the ascending node in the equatorial plane. Its range is defined as $0 \leq \Omega \leq 2\pi$.
(5) Argument of perigee $\omega$: the angle of rotation from the ascending node to the perigee along the direction of space target motion. Its range is defined as $0 \leq \omega \leq 2\pi$.
(6) True anomaly $\upsilon$: the angle of rotation from perigee to the current position of the target along the direction of the target motion. Its range is defined as $0 \leq \upsilon \leq 2\pi$. The true anomaly $\upsilon$, the mean anomaly $M$ and the eccentric anomaly $E$ are often used interchangeably according to the following conversion relation:

$$\begin{aligned} \tan\frac{v}{2} &= \sqrt{\frac{1+e}{1-e}}\tan\frac{E}{2} \\ M &= E - e\sin E \end{aligned} \tag{3.1}$$

When the initial conditions are known, the orbital elements of the space target at any time can be determined and have an explicit geometric meaning. The semi-major axis and eccentricity can determine the orbital characteristics (shape and size, as well as the related moment of momentum and energy) of the target. The orbital inclination and the right ascension of ascending node (RAAN) determine the orientation of the orbital plane in space. The argument of perigee determines the position of the apse line in the orbital plane. The true anomaly is a time parameter, and is the time reference of the space target moving along the orbit.

In general, the above variables can define the orbit of a space target and describe its position at any time except in the following two special cases:

(1) When e $\rightarrow$ 0 (i.e. nearly circular orbit), $\omega$ and $\upsilon$ are indefinite due to the uncertainty of perigee. Therefore, new variables can be introduced as the orbital elements fitting any eccentricity problem:

$$\begin{cases} a \\ i \\ \Omega \\ h = e\cos\omega \\ k = -e\sin\omega \\ L = v + \omega \end{cases} \tag{3.2}$$

(2) When $i \rightarrow 0°$ or $i \rightarrow 180°$, $\Omega$ and $\omega$ are indefinite due to the uncertainty of the ascending node, so new variables can be introduced:

$$\begin{cases} a \\ p = \sin i \cos \Omega \\ q = -\sin i \sin \Omega \\ h = e \cos(\omega + \Omega) \\ k = -e \sin(\omega + \Omega) \\ L = v + \omega + \Omega \end{cases} \tag{3.3}$$

or

$$\begin{cases} a \\ p = \tan \frac{i}{2} \sin \Omega \\ q = \tan \frac{i}{2} \cos \Omega \\ h = e \cos(\omega \pm \Omega) \\ k = -e \sin(\omega \pm \Omega) \\ L = v + \omega + \Omega \end{cases} \tag{3.4}$$

where the "±" sign in the expressions of $h$ and $k$ will be "±" when the orbit is prograde and will be "−" when the orbit is retrograde. The $h$ and $k$ in Eqs. (3.2) to (3.4) jointly describe the shape of the orbit and the position of the periapsis; the $p$ and $q$ jointly describe the direction of the orbital plane.

### 3.1.2 Orbit Prediction

The orbit prediction for a space target is to calculate its motion state or orbital elements at any time according to its initial orbital elements and the knowledge on orbit dynamics.

The space target is affected by not only the Earth's gravity but also various perturbations. To predict its orbit, we need to study its motion laws under various perturbations. The basic equation of the target motion [1] can be written as

$$\ddot{\boldsymbol{r}} = \boldsymbol{F}_0 + \boldsymbol{F}_e \tag{3.5}$$

where $r$ represents the position vector from the Earth's core to the target; $F_0$ is the gravitational acceleration of a particle of the Earth, i.e.

$$\boldsymbol{F}_0 = -\frac{\boldsymbol{\mu}}{r^2}\left(\frac{\boldsymbol{r}}{r}\right) \tag{3.6}$$

where $\mu$ represents the universal gravitational constant; $r$ represents the distance between the Earth's core and the target's center of mass; and $F_\varepsilon$ represent the force factors affecting the motion of the space target, for example, the Earth's oblateness, atmospheric drag, solar and lunar gravity, solar radiation pressure and other perturbations, and can be expressed as

$$\boldsymbol{F}_s = \sum_{k=1}^{n} \boldsymbol{F}_k(r, \dot{r}, t; \varepsilon) \tag{3.7}$$

where $t$ is time; $\varepsilon$ is a parameter, $\varepsilon \leq 1$.

$F_0$ and $F_\varepsilon$ satisfy the following condition:

$$|\boldsymbol{F}_s|/|\boldsymbol{F}_0| = O\left(\varepsilon^k\right) \tag{3.8}$$

where $O(\varepsilon^k)$ represents the $k$-order small quantity of $\varepsilon$. For different space targets in different mechanical environments, different reference values can be used as the small quantity.

The perturbation force (acceleration) is expressed in two forms. The first form is a right-handed helix system composed of the tangential component $U$, the inner normal component $N$ of the orbital plane and the outer normal component $W$ of the orbital plane, where the tangential direction is the motion direction of the target. Some perturbation forces are conservative forces with the potential $R$. Accordingly, the second form is

$$\boldsymbol{F}_s = gradR \tag{3.9}$$

where $R$ is also called perturbation function.

$U$-$N$-$W$ perturbation motion equation, also known as Guass-type perturbation motion equation, is expressed as

$$\begin{cases} \dfrac{\mathrm{d}a}{\mathrm{d}t} = \dfrac{2}{n\sqrt{1-e^2}}\left(1 + 2e\cos\nu + e^2\right)^{1/2} U \\ \dfrac{\mathrm{d}e}{\mathrm{d}t} = \dfrac{\sqrt{1-e^2}}{na}\left(1 + 2e\cos\nu + e^2\right)^{-1/2}\left[2(\cos\nu + e)U - \sqrt{1-e^2}\sin EN\right] \\ \dfrac{\mathrm{d}i}{\mathrm{d}t} = \dfrac{r\cos u}{na^2\sqrt{1-e^2}W} \\ \dfrac{\mathrm{d}\Omega}{\mathrm{d}t} = \dfrac{r\sin u}{na^2\sqrt{1-e^2}\sin i} W \\ \dfrac{\mathrm{d}\omega}{\mathrm{d}t} = \dfrac{\sqrt{1-e^2}}{nae}\left(1 + 2e\cos\nu + e^2\right)^{-1/2}[2\sin\nu U + (\cos E + e)N] - \cos i\dfrac{\mathrm{d}\Omega}{\mathrm{d}t} \\ \dfrac{\mathrm{d}M}{\mathrm{d}t} = n - \dfrac{1-e^2}{nae}\left(1 + 2e\cos\nu + e^2\right)^{-1/2} \\ \qquad \left[\left(2\sin\nu + \dfrac{2e^2}{\sqrt{1-e^2}}\sin E\right)U + (\cos E - e)N\right] \end{cases} \tag{3.10}$$

where $n = \sqrt{\mu/a^3}$ is the average angular rate of target motion, and $u = \upsilon + \omega$.

The $\partial R/\partial \sigma$ perturbation motion equation, also called Lagrange perturbation motion equation, is expressed as

$$\begin{cases} \frac{\mathrm{d}a}{\mathrm{d}t} = \frac{2}{na}\frac{\partial R}{\partial M} \\ \frac{\mathrm{d}e}{\mathrm{d}t} = \frac{1-e^2}{na^2 e}\frac{\partial R}{\partial M} - \frac{\sqrt{1-e^2}}{na^2 e}\frac{\partial R}{\partial \omega} \\ \frac{\mathrm{d}i}{\mathrm{d}t} = \frac{1}{na^2\sqrt{1-e^2}\sin i}\left(\cos i\frac{\partial R}{\partial \omega} - \frac{\partial R}{\partial \Omega}\right) \\ \frac{\mathrm{d}\Omega}{\mathrm{d}t} = \frac{1}{na^2\sqrt{1-e^2}\sin i}\frac{\partial R}{\partial i} \\ \frac{\mathrm{d}\omega}{\mathrm{d}t} = \frac{\sqrt{1-e^2}}{na^2 e}\frac{\partial R}{\partial e} - \cos i\frac{\mathrm{d}\Omega}{\mathrm{d}t} \\ \frac{\mathrm{d}M}{\mathrm{d}t} = n - \frac{1-e^2}{na^2 e}\frac{\partial R}{\partial e} - \frac{2}{na}\frac{\partial R}{\partial a} \end{cases} \tag{3.11}$$

Equations (3.10) and (3.11) contain the factors $\frac{1}{e}$ and $\frac{1}{\sin i}$ on their right-handed sides and thus cannot be used in the cases of $e = 0$ and $\sin i = 0$ (i.e. $i = 0°$ or $i = 180°$). Therefore, new variables should be introduced, as shown below:

$$\begin{cases} a \\ h = \sin\frac{i}{2}\cos\Omega \\ k = \sin\frac{i}{2}\sin\Omega \\ \xi = e\cos(\omega+\Omega) \\ \eta = e\sin(\omega+\Omega) \\ \lambda = M + \omega + \Omega \end{cases} \tag{3.12}$$

Accordingly, we obtain

$$
\begin{cases}
\dfrac{\mathrm{d}a}{\mathrm{d}t} = \dfrac{2}{na}\dfrac{\partial R}{\partial \lambda} \\
\dfrac{\mathrm{d}h}{\mathrm{d}t} = \dfrac{\cos i}{na^2\sqrt{1-e^2}}\left\{\dfrac{1}{2k+2h-1-\cos i}\left[-2h\dfrac{\partial R}{\partial \lambda} - (2h-1-\cos i)\dfrac{\partial R}{\partial k}\right]\right. \\
\qquad\qquad \left. + \dfrac{h}{1+\cos i}\left(\eta\dfrac{\partial R}{\partial \xi} - \xi\dfrac{\partial R}{\partial \eta} - \dfrac{\partial R}{\partial \lambda}\right)\right\} \\
\dfrac{\mathrm{d}k}{\mathrm{d}t} = \dfrac{\cos i}{na^2\sqrt{1-e^2}}\left\{\dfrac{1}{2k+2h-1-\cos i}\left[2k\dfrac{\partial R}{\partial k} + (2h-1-\cos i)\dfrac{\partial R}{\partial h}\right]\right. \\
\qquad\qquad \left. + \dfrac{h}{1+\cos i}\left(\eta\dfrac{\partial R}{\partial \xi} - \xi\dfrac{\partial R}{\partial \eta} - \dfrac{\partial R}{\partial \lambda}\right)\right\} \\
\dfrac{\mathrm{d}\xi}{\mathrm{d}t} = -\dfrac{2\eta\cos i}{na^2\sqrt{1-e^2}}\left(h\dfrac{\partial R}{\partial h} + k\dfrac{\partial R}{\partial k}\right) - \dfrac{\sqrt{1-e^2}}{na^2}\dfrac{\partial R}{\partial \eta} - \dfrac{\xi\sqrt{1-e^2}}{na^2\left(1+\sqrt{1-e^2}\right)}\dfrac{\partial R}{\partial \lambda} \\
\dfrac{\mathrm{d}\eta}{\mathrm{d}t} = \dfrac{2\xi\cos i}{na^2\sqrt{1-e^2}}\left(h\dfrac{\partial R}{\partial h} + k\dfrac{\partial R}{\partial k}\right) + \dfrac{\sqrt{1-e^2}}{na^2}\dfrac{\partial R}{\partial \xi} - \dfrac{\eta\sqrt{1-e^2}}{na^2\left(1+\sqrt{1-e^2}\right)}\dfrac{\partial R}{\partial \lambda} \\
\dfrac{\mathrm{d}\lambda}{\mathrm{d}t} = n - \dfrac{2}{na}\dfrac{\partial R}{\partial \lambda} + \dfrac{2\cos i}{na^2\sqrt{1-e^2}(1+\cos i)}\left(h\dfrac{\partial R}{\partial h} + k\dfrac{\partial R}{\partial k}\right) \\
\qquad\qquad + \dfrac{\sqrt{1-e^2}}{na^2\left(1+\sqrt{1-e^2}\right)}\left(\xi\dfrac{\partial R}{\partial \xi} + \eta\dfrac{\partial R}{\partial \eta}\right)
\end{cases}
\tag{3.13}
$$

And

$$
\begin{cases}
\dfrac{\mathrm{d}a}{\mathrm{d}t} = \dfrac{2}{na}\dfrac{\partial R}{\partial \lambda} \\
\dfrac{\mathrm{d}h}{\mathrm{d}t} = \dfrac{1}{2}\cos\dfrac{i}{2}\cos\Omega\dfrac{\mathrm{d}i}{\mathrm{d}t} - \sin\dfrac{i}{2}\sin\Omega\dfrac{\mathrm{d}\Omega}{\mathrm{d}t} \\
\dfrac{\mathrm{d}k}{\mathrm{d}t} = \dfrac{1}{2}\cos\dfrac{i}{2}\sin\Omega\dfrac{\mathrm{d}i}{\mathrm{d}t} + \sin\dfrac{i}{2}\cos\Omega\dfrac{\mathrm{d}\Omega}{\mathrm{d}t} \\
\dfrac{\mathrm{d}\xi}{\mathrm{d}t} = \cos(\omega+\Omega)\dfrac{\mathrm{d}e}{\mathrm{d}t} - e\sin(\omega+\Omega)\left(\dfrac{\mathrm{d}\omega}{\mathrm{d}t} + \dfrac{\mathrm{d}\Omega}{\mathrm{d}t}\right) \\
\dfrac{\mathrm{d}\eta}{\mathrm{d}t} = \sin(\omega+\Omega)\dfrac{\mathrm{d}e}{\mathrm{d}t} + e\cos(\omega+\Omega)\left(\dfrac{\mathrm{d}\omega}{\mathrm{d}t} + \dfrac{\mathrm{d}\Omega}{\mathrm{d}t}\right) \\
\dfrac{\mathrm{d}\lambda}{\mathrm{d}t} = \dfrac{\mathrm{d}\Omega}{\mathrm{d}t} + \dfrac{\mathrm{d}\omega}{\mathrm{d}t} + \dfrac{\mathrm{d}M}{\mathrm{d}t}
\end{cases}
\tag{3.14}
$$

By substituting this equation into Eq. (3.11), the singularities in Eq. (3.11) can be eliminated.

According to the above perturbation motion equation, the motion state or orbital elements of the space target at any time can be solved when the initial conditions are known. However, due to the complexity of the perturbation motion equation, only an approximate solution, rather than an exact analytical solution, can be obtained. Even if a high-precision solution can be given for some perturbation sources, the structure of the solution will be too complex to be specifically applied. Therefore, the numerical method is often used in orbit prediction. When the calculated arc is not too long, the numerical solution can easily meet the requirement of high precision.

### 3.1.3 Orbit Perturbation

The deviation of the actual motion of a space target relative to its motion in an ideal orbit (namely Kepler orbit) is called orbit perturbation. The orbit perturbation is caused by the following factors:

(1) The Earth is not uniformly spherical, so its gravitational acceleration cannot be accurately described in the form of $-(\mu/r^3)r$.
(2) In the space where the target is moving, thin air is still present and has an aerodynamic effect on the moving target. The lower the orbital altitude, the more obvious this effect.
(3) The sun and the moon also produce gravity on the space target. The farther the target is from the Earth, the more significant the relative effect of the solar and lunar gravity will be.
(4) The effect of solar radiation pressure.

According to their properties, the perturbation forces can be divided into conservative perturbation forces and non-conservative perturbation forces. The conservative perturbation forces (such as gravitational force) depend only on the position and can be described by a perturbative potential function. This type of perturbation force does not change the total energy of the space target and therefore will not change the semi-major axis of the orbit [2].

#### 3.1.3.1 Earth Oblateness Perturbation

The Earth's oblateness $J_2$ is the main perturbation term of space debris orbit, and its perturbation function is

$$R = -\frac{J_2 \mu R_E^2}{2a^3}\left(\frac{a}{r}\right)^3\left[\left(1 - \frac{3}{2}\sin^2 i\right) + \frac{3}{2}\sin^2 i \cos 2(v + \omega)\right] \tag{3.15}$$

where $R_e$ represents the radius of the Earth, and $J_2 = 1.0826300 \times 10^{-3}$.

The perturbation terms include a long period term and a short period term. The long period term will accumulate after integration, while the short period term will not accumulate after integration and will oscillate periodically about the original magnitude. For simplicity, the short periodic term that does not accumulate after integration can be separated from the perturbation function. Then the mean integral of the perturbation function $R$ within the orbital period $T$ of space debris is calculated to obtain the long period term (the $J_2$ perturbation function does not include the long period term):

$$R_e = \frac{1}{T}\int_0^T R\mathrm{d}t = -\mu R_E^2 \frac{J_2}{a^3}\left(1 - \frac{3}{2}\sin^2 i\right)\left(1 - e^2\right)^{-3/2} \tag{3.16}$$

Then the short period term will be

$$R_s = R - R_e \tag{3.17}$$

While the effect of short period is ignored, the effect of long term perturbation is mainly considered and substituted into Lagrangian perturbation motion equation to obtain:

$$\begin{cases} \dfrac{\mathrm{d}a}{\mathrm{d}t} = 0 \\ \dfrac{\mathrm{d}e}{\mathrm{d}t} = 0 \\ \dfrac{\mathrm{d}i}{\mathrm{d}t} = 0 \\ \dfrac{\mathrm{d}\Omega}{\mathrm{d}t} = -\dfrac{3}{2}\dfrac{J_2 R_E^2}{p^2} n \cos i \\ \dfrac{\mathrm{d}\omega}{\mathrm{d}t} = \dfrac{3}{2}\dfrac{J_2 R_E^2 n}{p^2}\left(2 - \dfrac{5}{2}\sin^2 i\right) \\ \dfrac{\mathrm{d}M}{\mathrm{d}t} = \dfrac{3}{2}\dfrac{J_2 R_E^2 n}{p^2}\left(1 - \dfrac{3}{2}\sin^2 i\right)\sqrt{1 - e^2} \end{cases} \tag{3.18}$$

As can be seen from Eq. (3.18), the long period effect of the Earth oblateness ($J_2$) perturbation does not change the semi-major axis, eccentricity and inclination of space debris orbit, but causes uniform changes in the right ascension of ascending node and the argument of perigee.

#### 3.1.3.2 Atmospheric Drag Perturbation

The aerodynamic force on space debris in the atmosphere is mainly manifested as a drag, and can be approximately expressed as

$$\begin{aligned} \boldsymbol{D} &= -\frac{1}{2}(C_D S)\rho V^2\left(\frac{\boldsymbol{V}}{V}\right) \\ \boldsymbol{V} &= \boldsymbol{v} - \boldsymbol{v}_a \end{aligned} \tag{3.19}$$

where $v$ and $v_a$ are the velocity vectors of space debris and atmosphere relative to geocentric coordinate system respectively; $S$ is the cross-sectional area of the spacecraft; $\rho$ is the atmospheric density. The terms in Eq. (3.19) are described below:

(1) The drag coefficient $C_D$ is related to the shape, attitude and orbital altitude of space debris, and is usually assumed to be $C_D = 2.2$.
(2) The atmospheric drag is a surface force, and is related to the cross-sectional debris area $S$ bearing the drag. The drag acceleration is related to the area-mass ratio $S/m$, and depends on the shape and attitude of the debris. In the absence of the information about debris shape and attitude, an equivalent area-mass ratio, namely $S/m =$ constant, can be adopted.
(3) In terms of atmospheric motion velocity $v_a$, atmospheric motion is usually manifested as a rotational motion. Its rotational angular velocity $\omega_a$ is more complicated, with a value range of $\omega_a = (0.8\text{–}1.4)n_e$, where $n_e$ is the rotational angular velocity of the Earth. For the LEO debris to which atmospheric drag perturbation is more important, its flying altitude is $h = 200\text{–}400$ km, so $\omega_a = n_e$ is accepted. As the orbital altitude increases, the effect of atmospheric drag will decrease, so $\omega_a = n_e$ is still accepted for low- and medium-orbit debris.
(4) In terms of the atmospheric density model $\rho = \rho(r, t)$, atmospheric density will decrease exponentially with the increase of orbital altitude.

To simplify the calculation, atmospheric rotation is not considered and then atmospheric drag acceleration can be written as

$$f_T = -\frac{1}{2}\frac{C_D S}{m}\rho\frac{\mu\left(1 + 2e\cos v + e^2\right)}{a\left(1 - e^2\right)} \tag{3.20}$$

By substituting the Eq. (3.20) into $U$-$N$-$W$ perturbation motion equation, we can obtain:

$$\begin{cases} \frac{da}{dt} = -\frac{\delta n a^2}{\left(1-e^2\right)^{3/2}}\left(1 + 2e\cos v + e^2\right)^{3/2}\rho \\ \frac{de}{dt} = -\frac{\delta n a}{\left(1-e^2\right)^{1/2}}\left(1 + 2e\cos v + e^2\right)^{1/2}(\cos v + e)\rho \\ \frac{di}{dt} = 0 \\ \frac{d\Omega}{dt} = 0 \\ \frac{d\omega}{dt} = -\frac{\delta n a^2}{e\left(1-e^2\right)^{1/2}}\sin v\left(1 + 2e\cos v + e^2\right)^{1/2}\rho \\ \frac{dM}{dt} = n + \frac{\delta n a}{e\left(1-e^2\right)}\left(\frac{r}{a}\right)\sin v\left(1 + 2e\cos v + e^2\right)^{1/2}\left(1 + e\cos v + e^2\right)\rho \end{cases} \tag{3.21}$$

where $\delta = \frac{C_D S}{m}$. On this basis, the rate of change of perihelion distance, $r_p = a(1 - e)$, can also be given:

$$\frac{dr_p}{dt} = (1 - e)\frac{da}{dt} - a\frac{de}{dt} \tag{3.22}$$

According to Eqs. (3.21) and (3.22), the semi-major axis and perihelion distance of space debris are decreasing under the atmospheric drag perturbation. It can known from the average effect that the eccentricity of the debris is also decreasing. Therefore, under the effect of atmospheric drag perturbation, the orbit of space debris will become smaller and more round, and finally settle and fall into the dense atmosphere. This is just the damping effect, which is manifested as the dissipation of orbital energy. The dissipation effect decides the life length of LEO debris orbit.

#### 3.1.3.3 Solar and Lunar Gravitational Perturbation

The solar and lunar gravitational perturbation accelerations $a_S$ and $a_M$ are the vector differences between the solar and lunar gravitational accelerations on space debris and the gravitational acceleration of the Earth, and can be expressed as

$$\begin{aligned} \boldsymbol{a}_S &= -\mu_S\left(\frac{\boldsymbol{r}-\boldsymbol{r}_S}{|\boldsymbol{r}-\boldsymbol{r}_S|^3}+\frac{\boldsymbol{r}_S}{|\boldsymbol{r}_S|^3}\right) \\ \boldsymbol{a}_M &= -\mu_M\left(\frac{\boldsymbol{r}-\boldsymbol{r}_M}{|\boldsymbol{r}-\boldsymbol{r}_M|^3}+\frac{\boldsymbol{r}_M}{|\boldsymbol{r}_M|^3}\right) \end{aligned} \tag{3.23}$$

where $\mu_S$ and $\mu_M$ are the solar and lunar gravitational constants respectively; $r_S$ and $r_M$ are the position vectors of the sun and the moon in J2000 inertial coordinate system respectively; $r$ is the position vector of space debris in J2000 inertial coordinate system. The lower the orbit of space debris, the smaller the solar and lunar perturbation force on the debris. However, for high-orbit debris, the perturbation force is very significant. At the altitude above 50,000 km, the effect of solar and lunar gravitational perturbation outweighs the oblateness of the Earth. The actual calculation shows that the solar and lunar gravitational perturbation mainly depends on the shape of debris orbit, the position of the orbital plane, as well as the positions of the apse line relative to the lines between the moon and the Earth and between the sun and the Earth. If these connecting lines are located in the orbital plane of the debris, the solar and lunar gravitational perturbation will not affect the orbital plane position. For a circular or nearly circular orbit, the orbit is always relatively stable as long as the inclination of the orbital plane relative to the ecliptic plane is small. However, for an elliptical orbit, if the apogee of the debris orbit is outside the lunar orbit, the perigee altitude of the debris orbit will drop.

#### 3.1.3.4 Solar Radiation Pressure Perturbation

Solar radiation pressure is the force generated when the photon flow of solar radiation collides with the surface of space debris. The debris outside the Earth's shadow will be exposed to this solar radiation pressure. The acceleration of solar pressure perturbation can be expressed as

$$a_{sr} = k_v C_R \frac{S}{m} p_{sr} \mathrm{AU}^2 \frac{\boldsymbol{r} - \boldsymbol{r}_S}{|\boldsymbol{r} - \boldsymbol{r}_S|} \tag{3.24}$$

where $C_R$ is the solar pressure parameter related to debris surface material, whose dimension is 1; $p_{sr}$ is the solar radiation constant, and can be approximated as $4.56 \times 10^{-6}\mathrm{N{\cdot}m^{-2}}$; AU is the mean distance between the sun and the Earth, AU = 1.495979 $\times 10^{11}$ m; $k_v$ is the shadow coefficient ($k_v = 0$ when the debris is in the umbra of the Earth; $k_v = 1$ when the debris is completely in sunlight; $0 < k_v < 1$ when the debris is in the penumbra of the Earth).

As can be seen from Eq. (3.24), like atmospheric drag perturbation, radiation pressure perturbation is also related to the area-mass ratio of space debris. When the orbital altitude of the debris is low, the effect of atmospheric drag perturbation cannot be ignored as a dissipative force that can reduce the motion energy of the debris (namely, the semi-major axis decreases monotonically with the increase of time). However, when the orbital altitude of the debris is high, the effect of solar pressure perturbation will outperform the atmospheric drag perturbation.

## 3.2 Space Environment Model

Since the initial Meteoroid/Orbital Debris (M/OD) environment modeling at the NASA Johnson Space Center (JSC) in the late 1970s, significant research progress has been made in the field of M/OD environment modeling through nearly 40 years of continuous joint efforts of space powers. A series of M/OD environment models have been established, providing a important technical support for the environment forecasting, risk assessment and protection design of M/OD. The M/OD environment models built upon the data of M/OD detection, ground-based hypervelocity collision simulation test and numerical simulation can be roughly classified into short-term engineering models and long-term evolution models according to their different modeling methods and application scopes. The short-term engineering models are semi-empirical models mainly based on various types of detection data. The effective period of these models is generally not more than 10 years because they do not take into account the influence of such factors as the source, evolution and extinction of space debris. However, new detection data can be used to correct and update the short-term models, mainly to provide the current M/OD environment data for M/OD risk assessment and protection design. The long-term evolution models are semi-deterministic models. In addition to all types of detection data, they also take into account the influence of two types of factors, namely the "sources" of debris (e.g. launching, on-orbit operation, collision fragmentation and on-orbit explosion) and the "die-outs" of debris (e.g. atmospheric drag, solar and lunar perturbation, and solar radiation pressure). Therefore, the long-term models have a longer effective period. They are mainly used to evaluate the current situation of M/OD environment and forecast its future evolution behavior. Typical foreign M/OD environment models are listed in Table 3.1.

**Table 3.1** Space debris models

| Model name | Ownership country and organization | Type | Min. debris size/ mm | Applicable orbits |
|---|---|---|---|---|
| ORDEM | NASA | Short period | 0.01 | LEO/GEO |
| CHAIN | NASA | Long period | 10 | LEO |
| EVOLVE | NASA | Short and long periods | 1 | LEO |
| LEGEND | NASA | Short and long periods | 1 | LEO to GEO |
| GEOEVOL VE1.0 | NASA | Short and long periods | 300 | GEO |
| MASTER | ESA | Short period | 0.001 | LEO/GEO |
| CHAINEE | ESA | Long period | 10 | LEO |
| NAZARENKO | RSA | Long period | 0.6 | LEO |
| SPDA | RSA | Short and long periods | 1 | LEO/GEO |
| IDES | Defense Evaluation and research agency (UK) | Short and long periods | 0.01 | LEO |
| LUCA | Technical University of Braunschweig (Germany) | Long period | 1 | LEO/MEO |
| SDM/STAT | ESA/Italian Space Agency | Short and long periods | – | LEO/GEO |

### *3.2.1 Short-Term Engineering Model*

The short-term engineering models are mainly established for debris protection design in spacecraft engineering. Although they may not directly reflect the sources of space debris, their output M/OD flux on spacecraft surface is required to be consistent with the measured result. The following are the modeling methods, principles and characteristics of short-term engineering models developed by NASA and ESA and widely used at home and abroad.

#### 3.2.1.1 NASA89/91 Models

NASA89 model is the first short-term engineering model of space debris environment established by NASA. In 1991, the NASA91 engineering model was established by expanding and updating the NASA89 data sources. NASA89 and NASA91 are also the first versions of later ORDEM models, and are used by NASA to describe the space debris environment in the LEO region, especially at the orbital altitudes below 1000 km.

In the NASA89 model, the space debris flux is directly described and given as a simple function. Compared with NASA89 model, the NASA91 model has no substantial change in the debris environment characterization, but contains some adaptive modifications of the debris flux function that were made with the accumulation and enriching of detection data.

The space debris flux on the spacecraft surface can be directly calculated using a model proposed by NASA. For example, for an arbitrarily flipped spacecraft launched in 2000 into an orbit with an inclination of 30° and an altitude of 500 km, the calculation results of debris flux distribution on the spacecraft surface are shown in Figs. 3.1 and 3.2. As can be seen from Fig. 3.1, for the debris with a diameter of $d < 1$ cm, the debris flux estimation given by NASA91 model is smaller than that given by NASA89 model, mainly because the two models adopt different debris growth rates. In NASA91 model, the annual growth rate of small debris is assumed to decrease from 10 to 2%. However, for the debris with a diameter of $d > 1$ cm, the flux estimation given by NASA91 model is larger than that given by NASA89 model, mainly because the flux correction function $H(d)$ of NASA91 model is greater than 1. As can be seen from Fig. 3.2, the debris flux increases with the orbital altitude, but basically remains unchanged when the altitude exceeds 800 km. This is because the NASA91 model is only applicable to the orbital altitudes below 1000 km, and the amount of small debris is much larger than that of large debris at the same orbital altitude.

**Fig. 3.1** Estimation of debris flux distribution versus size by different models

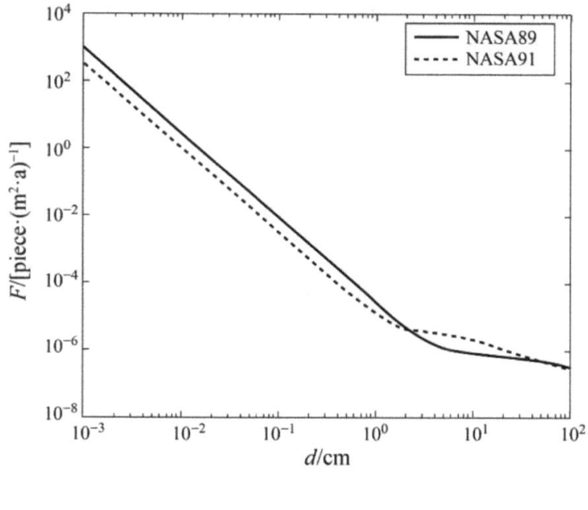

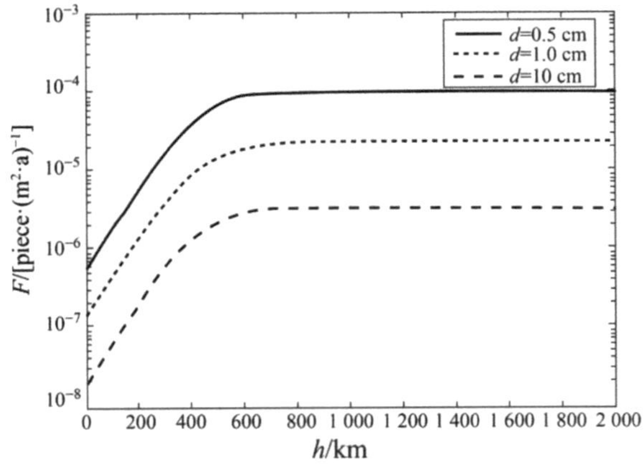


**Fig. 3.2** Flux distribution of space debris of different sizes versus orbital altitude

#### 3.2.1.2 ORDEM96 Model

Different from NASA89 and NASA91 models, the ORDEM96 model does not directly give the debris flux distribution on the spacecraft surface, but calculates the flux distribution by describing the orbital elements of LEO debris. The effectiveness of ORDEM96 has been verified by the Haystack radar observation data used for American satellite cataloging, the recovery analysis of the long duration exposure facility (LDEF), and the post-mission inspection of the US space shuttle.

In the ORDEM96 model, the LEO region is divided into 6 different orbital inclination zones (i.e. 7°, 28°, 51°, 65°, 82° and 98°) and 2 eccentricity groups (i.e. circular orbit and highly-elliptical orbit). Then, a representative orbit is selected from each inclination zone to calculate the distribution of fluxes, angular fluxes and velocities of the debris on the spacecraft surface and in the field of view of a ground-based detector.

In the ORDEM96 model, the space debris is divided by size into six categories, namely complete objects (defunct satellites, defunct rocket bodies, etc.), large fragments, small fragments, sodium–potassium particles, coating fragments, and alumina. By processing their data, the distribution of surface fluxes, impact fluxes or breakdown fluxes along a certain direction can be obtained.

#### 3.2.1.3 ORDEM2000 Model

As an updated version of ORDEM96 model, the ORDEM2000 model was officially released in 2000, with a validity period from 1991 to 2030. It is suitable for the analysis of 10 pm–1 m space debris environment in the LEO region. Compared with NASA89/NASA91 and ORDEM96, the ORDEM2000 model has been significantly improved in terms of data source, modeling method, result output, analysis accuracy, program function and interactive interface, and is an important breakthrough in the field of space debris environment modeling.

#### 3.2.1.4 ORDEM2008 Model

ORDEM2008 model is the latest short-term engineering model of space debris environment released by NASA in 2008. It is a further improvement and update of ORDEM2000 model, applying to the orbital altitude range expanded from LEO region to GEO region, that is, 200–3600 km. From the perspective of modeling method, there is no essential difference between ORDEM2008 model and ORDEM2000 model. However, due to the introduction of various advanced detection means and the further improvement and enrichment of debris data sources, the functions that could not be implemented under the framework of ORDEM2000 model have been expanded and improved in ORDEM2008 model in recent years. Thus, the ORDEM2008 model has a higher application value, richer output data and more friendly use. The main differences between the ORDEM2000 and ORDEM2008 models are listed in Table 3.2. The data sources of the ORDEM2008 model are listed in Table 3.3.

As can be seen from Table 3.2, the ORDEM2008 model not only divides the debris sizes in more detail by adding five sizes ($\geq 31.6\ \mu m$, $\geq 316\ \mu m$, $\geq 3.16$ mm, $\geq 3.16$ cm and $\geq 31.6$ cm) to the original sizes, but also classifies the debris sources, debris densities and applicable orbital altitudes, thus obtaining more detailed analysis results and richer output data. In addition, the ORDEM2008 model has improved the graphical user interface, allowing users to gain more detailed and richer graphical analysis results.

#### 3.2.1.5 MASTER Model

MASTER series model is a semi-deterministic model established by ESA based on the data of debris velocity/density distribution through three-dimensional discretization, applying to the short-term engineering prediction of M/OD environment in the region extending from LEO to GEO. The original MASTER model was MASTER95, but the first edition officially released by ESA was MASTER97. The current latest version is MASTER2005. Due to the limitation to the analysis of space debris distribution with a size of at least 0.1 mm and the lack of a graphical user interface, the MASTER95/MASTER97 models cannot be used for post-flight recovered-surface analysis and other studies.

Compared with MASTER97 model, the improved MASTER99 model (released in July 2000) has introduced a series of new debris sources such as the combustion slag of solid rocket motor (SRM), the NaK coolant released from Russian RORSAT satellite reactor, the surface degradation products and the surface degradation particles generated by radiation and atomic oxygen environments, the sputtered particles, and the spalling products generated by small collisions, while expanding the minimum

**Table 3.2** Main differences between ORDEM2000 and ORDEM2008

| Model name | ORDEM2000 | ORDEM2008 |
|---|---|---|
| Applicable time frame | 1991–2030 | 1995–2035 |
| Applicable orbital altitude and debris size / km | 200–2000 (≥ 10 μm) | 200–600 (≥ 10 μm)<br>600–2000 (≥ 1 mm)<br>2000–33,000 (≥ 1 cm)<br>34,000–38,000 (≥ 10 cm) |
| Debris source classification | N/A | Complete fragment<br>Low density fragment<br>Medium density exfoliation/splashing<br>High density fragment<br>Medium density alumina<br>NaK coolant |
| Debris density classification | N/A | Low density (< 2 $g/cm^3$)<br>Medium density (2–6 $g/cm^3$)<br>High density (> 6 $g/cm^3$)<br>NaK coolant (0.9 $g/cm^2$) |
| Debris size classification | ≥ 10 μm, ≥ 100 μm, ≥ 1 mm, ≥ 1 cm, ≥ 10 cm, ≥ 1 m | ≥ 10 μm, ≥ 31.6 μm, ≥ 100 μm, ≥ 316 μm, ≥ 1 mm, ≥ 3.16 mm, ≥ 1 cm, ≥ 3.16 cm, ≥ 10 cm, ≥ 31.6 cm, ≥ 1 m |

**Table 3.3** Data sources of ORDEM2008 model

| Means of acquisition | Application | Region |
|---|---|---|
| SSN cataloging | Data of complete fragments and large fragments | LEO > 10 cm, GEO > 70 cm |
| CobraDane radar | Comparison with SSN catalog data | LEO > 4 cm |
| Haystack radar | Fragmentation statistics | LEO > 1 cm |
| Goldstone radar | Comparison with Haystack detection data | LEO > 2 mm |
| STS porthole and radiator recovery surface | Fragmentation statistics | LEO < 1 mm |

debris size to 1 m. In addition, the MASTER99 model adds the typical fragments smaller than 1 cm from the new debris sources to the cataloged objects, and then normalizes them into a reference debris source with a reference time of August 1, 1999, and finally determines the distribution of debris fluxes and debris densities by numerical integration.

Compared with MASTER99 model, the MASTER2001 model not only expands the data source, but also improves the graphical user interface, so as to implement the simulation in other tools through external data interface program.

MASTER2005 model is the latest version of MASTER series model released by ESA in 2005, and is the improvement and update of MASTER2001 model based on the accumulation and enrichment of detection data. However, the modeling method of MASTER2005 model is not essentially different from MASTER2001 model and is the same as ORDEN2000 model. In addition, by using the Program for Orbital Debris Environment Modeling (POEM) to update the debris database (including the generation, evolution and extinction of debris), the MASTER2005 model can obtain the debris data sources at different time and output richer data in a more user-friendly way. The application scopes and characteristics of MASTER models are listed in Table 3.4.

### *3.2.2 Long-Term Evolution Model*

Space debris evolution model is a long-term evolution model. The "long-term" here is relative to the short-term engineering model. The evolution model focuses more on describing the space debris environment from the perspective of mechanism and is mainly used to predict the long-term evolution behavior of the entire space debris environment in the future. During modeling, the current debris environment database should be built upon the existing detection data, and a series of supporting sub-models, such as launching activity model, orbital decay model and explosion event model, should be introduced in accordance with the impact of the self-growth and self-extinction of debris [3].

**Table 3.4** Application scopes and characteristics of MASTER models

| Model name | MASTER95 | MASTER97 | MASTER99 | MASTER2001 | MASTER2005 |
|---|---|---|---|---|---|
| Object size/μm | > 100 | 100 | > 1 | > 1 | > 1 |
| Orbital altitude | LEO-GEO | LEO-GEO | LEO-GEO | LEO-GEO | LEO-GEO |
| Time frame | 1995.03.30 ± 10a | 1996.12.31 ± 10a | 1999.08.01 ± 10a | 1960—2050 | 1957—2055 |
| Debris source | Fragments of disintegrating cataloged objects | Fragments of disintegrating cataloged objects | Fragments of disintegrating cataloged objects;<br>SRM slag;<br>NaK coolant;<br>Surface degradation products;<br>Sputtered particles | Fragments of disintegrating cataloged objects;<br>SRM slag;<br>NaK coolant;<br>Surface degradation products;<br>Sputtered particles;<br>West Ford copper needle | Fragments of disintegrating cataloged objects;<br>SRM slag;<br>NaK coolant;<br>Surface degradation products;<br>Sputtered particles;<br>West Ford copper needle |
| Micrometeoroid model | Grun | Divine-Staubach | Divine-Staubach | Divine-Staubach | Divine-Staubach (updated) |
| Model feature | Spherical flux<br>Arbitrary target orbit | Spherical flux<br>Arbitrary target orbit | Spherical flux;<br>Arbitrary target orbit;<br>Directional surface flux;<br>Spatial density;<br>GUI | Spherical flux;<br>Arbitrary spacecraft orbit;<br>Directional surface flux;<br>Spatial density;<br>GUI (updated);<br>Time browser | Spherical flux;<br>Arbitrary spacecraft orbit;<br>Directional surface flux;<br>Spatial density;<br>GUI (updated);<br>Time browser |

#### 3.2.2.1 Components of Long-Term Evolution Model Architecture

The modeling process of a space debris evolution model is to establish the current debris environment database with the existing detection data and then introduce a series of supporting sub-models, such as launching activity model, explosion event model, collision fragmentation event model, leakage event model, orbital decay model and mitigation model, in accordance with the impact of the "sources" (generation) and "die-outs" (extinction) of debris to calculate and predict the space debris environment at a specific time in the future. Among the sub-models, the launching model is used to describe the growth trend of debris in various orbits. The explosion model, fragmentation model and leakage model are used to describe the amount, size and orbital distribution of the debris generated by explosions, collisions and leakages. The orbital decay model is used to describe the changing of debris orbit parameters with time under various perturbation forces. The mitigation model is used to describe the impact of various mitigation standards and technical measures on the evolution of space debris environment.

1 **On-orbit Disintegration Model**

The disintegration of a spacecraft in orbit is the most important source of sudden debris increase. In particular, a large number of fragments of centimeter level and below generated by disintegration are difficult to track, determine (in terms of orbit) and catalogue, and can't be handled with effective protection methods and means. Thus they pose a great threat to the safe operation of other orbiting spacecrafts. On-orbit disintegration model is a semi-empirical estimation model that describes the spacecraft behavior of on-orbit explosion or collision fragmentation and the distribution characteristics of debris. The on-orbit disintegration models are classified into several categories: high-intensity explosion models, low-intensity explosion models and collision fragmentation models. According to the classification and description of on-orbit disintegration models in foreign literature, the disintegration behavior caused by the explosion of the remaining spaceborne propellant belongs to the high-intensity explosion disintegration, which is characterized by the generation of a large number of fragments after explosion and is the main source of small debris of centimeter level or below.

2 **Launch Event Model**

Space launch events are the main source of large space objects. Since 1957, more than 5500 spacecrafts have been put into orbit in more than 4300 launches. Since 1990, 70–80 launches have been made annually on average, for example, 75 launches in 1995 and 73 launches in 1996. The launch of large LEO constellations (such as the Motorola Iridium consisting of 66 working satellites, and the Teledesic consisting of 288 working satellites) may change this trend to some extent. In addition to the payload, multiple pieces of space debris (such as the final stage of a carrier rocket and the orbital module of a manned spacecraft) usually stay in orbit after each launch. The space launches can be roughly divided into three categories, namely: conventional launches, satellite constellation launches, and the launch activities related to the construction of large-sized space structures.

In consecutive time intervals, different launch rates may be chosen to simulate a changing launch policy (taking into account the inherent number of launches for a particular event in a given year, such as a sharp change in the number of launches due to repeated launcher failures). After calculating the total number of objects launched into orbit in one time interval, the orbital elements and the object masses are specified, and then the next time interval is calculated according to the statistics of past launches. In addition to conventional launches, satellite constellation launches should also be considered. The constellation construction, in which a large number of objects are launched into orbit in a short period of time, will significantly change the conventional launch activities. For example, if the constellation achieves the standard structure in the first year, the launch rate will be large that year, and the later launches will be only for the replacement of decommissioned or defunct satellites. In the model, the launch rates for two different periods can be defined. The orbital elements of the constellation, the number of spacecrafts remaining in space after each launch, the number of final-stage rocket bodies, and the number of mission-related fragments can be specified. Finally, the construction and maintenance of large space structures (such as the International Space Station) should be considered. In this case, the assembly and service missions can be predicted, but no new objects will be launched into orbit.

## 3 Leakage Event Model

According to the observation of American Haystack radar, in the circular orbit with an altitude of 600–1000 km and an inclination of about 65°, there are as many as 50,000–70,000 fragments larger than 8 mm, but very few fragments larger than 3 cm. The ballistic coefficient of the fragments is basically the same as that of particles with a density of about 1 g/cm$^3$. American and Russian experts believe that this particular distribution of debris is likely caused by a fuel leakage event and even dates back to the failure of Soviet "Cosmos 954" launch vehicle in 1978, and the leakage events of 16 RORSAT satellites from "Cosmos 1176" (launched in 1980) to "Cosmos 1932" (launched in 1988). Moreover, the distribution of liquid leakage particles obtained by the least square method is basically consistent with that of the 16 leakage events. Leakage event is a low-energy particle release behavior. In a leakage event, liquid particles usually enter space at very low speeds of a few meters per second. Historically, the typical leakage events were the leaking of liquid NaK coolant from the reactors of the Soviet RORSAT satellites.

## 4 Orbital Decay Model Considering Environmental Perturbation

In the LEO region (especially at the orbital altitudes of several hundred kilometers), space debris will gradually deviate from their original orbit under the influence of various perturbation forces, and will eventually enter the atmosphere and die out by itself after a long time of orbital decay. The perturbation forces mainly include the Earth's oblateness, atmospheric drag, solar radiation pressure, as well as solar and lunar gravity.

#### 3.2.2.2 Typical Long-Term Evolution Models

The EVOLVE, LEGEND and CHAIN/CHAINEE models are widely used to predict the evolution of space debris environment at a specific time in the future.

The EVOLVE4.0 model is a widely used model developed by NASA using the Monte Carlo stochastic simulation method. It is composed of four sub-modules: launch, disintegration, SRM combustion slag and reactor coolant leakage, and solar activity. These four sub-modules are equivalent to four characteristic quantities, which can predict the changes of orbital elements and debris flux and the probability of collisions between objects in space.

The LEGEND model also uses the Monte Carlo simulation method to calculate the evolution of all space targets at the altitudes of 200–50,000 km below the geosynchronous orbit (GSO). It includes the numerical calculation models of future launch events, atmospheric drag and solar radiation, while considering the historical launch model. The LEGEND model can give the debris distribution related to time, altitude, longitude and latitude, for example, the distribution of debris sizes, densities and velocities.

The CHAIN/CHAINEE model, which is based on the Particle-in-a-box model, simplifies the space debris environment into a series of random variables in different mass domains and orbital altitude domains, and proposes a new method for the analysis of a long-term evolution behavior. In contrast, the models like EVOLVE are generally not suitable for analyzing the evolution behavior of space debris involving continuous collisions due to a large consumption of computer memory. Through simplification, the CHAIN model has become $10^3$ to $10^4$ times faster than EVOLVE model, but with a lower confidence level. The CHAIN model is mainly used to analyze the long-term evolution and chained collisions of space debris.

#### 3.2.2.3 Basic Applications of Long-Term Evolution Models

The applications of long-term evolution models usually include the calculation of space debris life, the assessment of debris collision risks, and the mitigation of Kessler Syndrome.

1. **Calculation of Space Debris Life**

The estimation of space debris life is mainly aimed at the LEO debris, whose extinction is mainly caused by the perturbation forces, such as the Earth's non-spherical perturbation, atmospheric drag, solar radiation pressure, and solar and lunar gravity. The existing methods for debris life calculation mainly include numerical method, semi-analytical method and analytical method. The numerical method has high precision but long calculation time. The semi-analytic method has the computation speed greatly improved by using average orbital elements as the state variables. The key to debris life calculation modeling is the selection of an atmospheric model. However, it is often difficult to establish an accurate atmospheric model because atmospheric density is affected by many factors such as time and space. The current studies in this regard mainly focus on the falling of satellites into the atmosphere. Some of them have concluded through analysis that the factors affecting the prediction accuracy have a variation range of 15–20%. In fact, the most fuel-efficient way to kill a LEO spacecraft is to maneuver it into an orbit doomed to extinct in 25 years, usually at an altitude of less than 600 km. The space debris at this altitude will die out in 25 years.

2. **Debris Collision Risk Assessment**

The collision risk assessment model for space objects is mainly based on the collision probability model, which can be established with many methods. For example, a deterministic method based on geometric distance approaching events[1] is to detect the collision probability by altitude screening, orbit geometry screening and phase screening in succession. Generally speaking, the altitude screening is to screen out the space objects intersecting the altitude between the perigee and apogee of the target orbit, while taking into account the altitude attenuation and short period changes. Then, the orbit geometry screening is carried out for selected objects to determine the intersection of the orbital plane of a dangerous object with the orbital plane of the object of interest and to calculate the closest distance, which is then compared with the safe distance. Finally, the orbit phase screening is carried out to analyze whether the time windows of the two objects crossing the intersection will overlap in the forecast time. In addition, based on the idea of the collision probability increasing with spatial density, some scholars divide the space into grids and count the fragments in each grid at the sampling time by using the method of unified sampling. If the number of fragments in a grid is greater than or equal to 2, a certain probability of collision must exist between fragments.

3. **Kessler Syndrome Mitigation**

With the increase of space debris density, more and more people are wondering whether there is a limit of space debris density. If the limit is exceeded, the space exploration will have difficulty in taking a step. Intuitively, this limit must exist. The exceeding of this density limit will trigger a series of positive feedback phenomena, resulting in a collision cascade effect. As a result, the density will increase inexorably and thus render space unusable. The collision cascade effect was first studied by the NASA expert Donald J. Kessler, and thus was named after him as "Kessler Syndrome". Kessler raised the possibility of a vicious cycle, in which the increase of debris density at a certain orbital altitude to a critical level would result in more serious congestion in the orbit, a larger likelihood of collisions between debris, and then the generation of more new debris whose collisions, in turn, would make space even more congested.

According to the previous debris evolution models, the debris can die out in a variety of ways. In some people's view, Kessler's estimate is too pessimistic since the debris density will always be kept in a stable and controllable range given no satellite

[1] Liou J C. Orbital debris mitigation in support of space situational awareness and space traffic management, 2019.

collision and other behaviors in a time frame that can lead to the increase of debris amount. But in fact, Liou and Johnson used LEGEND model[2] in an evolutionary analysis of the space object environment for the next 200 years, assuming that the launch activities would cease after 2005. According to their projections, even if no new satellite collisions occurred from 2006 onwards, the number of explosion-generated fragments and the number of mission-related fragments in near-Earth orbit would go down, but the number of collision-generated fragments and the total number of fragments would go up in the next 200 years. Meanwhile, Liou and Johnson pointed out that the large number of analyses based on different assumptions and models all came to the same conclusion, that is, the amount of debris in low-Earth orbit (LEO) would increase inexorably, and that this unsafe situation would last for a long time [3]. Moreover, the cascading effect of such collisions has been identified in literature as the most important reason for the phenomenon that the number of fragments smaller than 1 mm has now exceeded 1 million.

## 3.3 Space Environment Analysis Under the New Situation

### *3.3.1 Impact of StarLink Program on Space Environment*

In recent ten years, as more and more companies began to propose the launch programs of large LEO constellations, the impact of large constellations on the future space debris environment has received wide attention. In particular, the Federal Communications Commission (FCC) approved SpaceX to deploy a constellation of 2825 satellites at an altitude of 1110–1325 km and another constellation composed of 1584 satellites at an altitude of 550 km and 7518 satellites at an altitude of 335–346 km. Together, the two constellations made up the SpaceX system. In October 2019, the SpaceX proposed to the FCC the addition of a constellation of 30,000 satellites, bringing the total number of constellation satellites to 42,000. The large number of satellites in these constellations pose new challenges to the entire space debris environment. Therefore, it is particularly important to properly manage the constellation launches and take reasonable measures to limit the impact of large constellations on the space debris environment. In this section, the "StarLink" program will be taken as an example to analyze the possible impact of large-scale constellations on space environment and even space security.

In January 2015, Musk, CEO of SpaceX, proposed the "StarLink" program. The program aims to launch 12,000 satellites into low-Earth orbit in the years 2019–2024, creating a giant three-tier satellite network. The three tiers, located at the orbits of 340, 550 and 1150 km respectively above the ground, will eventually connect all

[2] Liou J C. USA space debris environment, operations, and research updates [C]. 54th Session of the Scientific and Technical Subcommittee, Committee on the Peaceful Uses of Outer Space. Vienna: UNCO-POUS, 2017.

the satellites into a giant "constellation" that will provide a high-speed and low-cost all-weather satellite Internet to the entire Earth (including the north and south poles).

The SpaceX is expected to include about 42,000 satellites. According to SpaceX's report, these satellites are only used for building a global interconnected information network, but undoubtedly, the Starlink program has a huge application potential in the military field. Reportedly, the SpaceX leased some of its launch sites from the USAF Vandenberg Air Force Base so as to exempt some of its top-secret programs from scrutiny. Moreover, the early proof tests of Starlink's low-orbit technology demonstrated a direct link between a satellite and an antenna array aboard an USAF fighter. Thus, the Starlink program has a deep American military background, with the following further military ambitions[3]:

The Starlink program will help the United States seize the limited LEO spectrum resources. On the one hand, the LEO satellites have a more important military value than MEO and HEO satellites. They are generally less than 1000 km above the ground and thus have a smaller space transmission loss. The coverage area formed by LEO satellites moves faster across the Earth's surface, requiring only two hours to circle the Earth. This means that the use of LEO satellites can provide greater communication bandwidth, faster transmission speed and higher reconnaissance efficiency.

On the other hand, orbital frequency bands are limited natural resources in the aerospace field, and satellite companies need to apply for the use of these bands to relevant agencies in the way of declaration. According to the rules set by International Telecommunication Union (ITU), orbital resources are allocated mainly on a "first come, first served" basis, that is, the latter declarers are not allowed to adversely interfere with the satellites of the first declaring country. However, the LEO constellation system should not only launch its satellites in accordance with the ITU rules, but also provide relevant services to the public in accordance with the specified time and proportion, so as to legalize and fix its frequency status.

In other words, the development and utilization of a LEO constellation system can help seize not only the limited LEO satellite resources but also the spectrum initiative, which has an important impact on the coverage of battlefield communication channels. In addition, unlike the GEO satellites, which exist as a single satellite, the LEO satellites need to be networked in order to provide the communication services. A GEO satellite at EOL will rise by more than 200 km on its final energy and enter a "graveyard orbit" for "self-destruction", thereby giving up the orbit it occupies. In contrast, a LEO constellation is made up of many small satellites. When some of the satellites fail, the original launcher can launch new satellites to fill the vacancies, rather than giving up the entire constellation resource to others.

Therefore, the competition for resources is more intense in the field of LEO constellations. Compared with the development of HEO and MEO satellite network business, the emerging LEO communication satellite has become the main track for different countries in different camps to compete in space. The implementation of the "Starlink" program will further help the United States seize the limited space spectrum resources of LEO satellites.

---

[3] Aerospace Vision, the Military Ambition behind StarLink Program.

From the perspective of space warfare, the Starlink will also become the "space-based brain" of the USAF drone swarms. With the accelerated evolution of war form into intelligent war, the drone operation has become an important part of modern wars. In the early morning of January 3, 2020, the USAF MQ-9 "Reaper" drone accurately killed Qasem Soleimani, commander of Iran's elite special force "Quds Force", demonstrating the killing power of the drone supported by a powerful combat system. In order to cope with the future systematic and information-based battlefield environment featuring strong confrontation, the combat mode of drones has begun to develop from "single drone" operation to "swarm" operation. A large number of drones are carried and released by large- and medium-sized aerial platforms, forming a swarm that can launch wide-area distributed multi-point multi-direction attacks through self-organizing collaborative algorithm and pose a great challenge to traditional defense systems. In 2012, the US Navy made hundreds of simulated drone swarm attacks and found that even the Aegis-equipped destroyers were unable to cope with the swarm attacks. When the swarms, each consisting of eight drones, made attacks, 2.8 drones on average were able to evade interception. Even when the ship-based air defense system was upgraded, at least one drone was still able to evade interception. If the number of drones increases to dozens or even hundreds, the result will be self-evident.

Although the drone swarm operation has many advantages, the real-time information exchange between swarms is needed to achieve collaborative command and control through "Achilles Heel", the communication link and control system. For example, 103 "Grey Partridge" micro-drones were released from three USAF F/A-18F "Super Hornet" fighters, demonstrating the cooperative combat capabilities such as collective decision-making and adaptive formation flight. The core support was the sharing "Combat Cloud" (a distributed data interaction server) that communicated with multiple ground stations. Once its communication links were interfered with or its control system was hacked, the whole drone swarm would be likely wiped out.

In other words, the overall survivability of a swarm depends largely on its communication control ability. The "Starlink" program can effectively fill this gap. The large number of LEO satellites in "Starlink" have large communication bandwidth, low delay and global coverage, and are insensitive to any terrain (such as mountain, ocean or polar region) or weather. The "Starlink" can free all kinds of low-altitude aircraft from their dependence on land-based communication systems and serve as the "space-based brain" of drone swarms in warfare. In addition, through flight control, situational awareness, information sharing, target allocation and intelligent decision-making, the "Starlink" can expand the capacity of a single drone and significantly improve the overall combat effectiveness of a whole swarm.

### 3.3.2 Research Status of Large-Scale Constellations

With the great progress made in a series of spaceflight technologies in recent years, the threshold to enter and utilize space has been greatly lowered, and the commercial value of spaceflight has become increasingly prominent and attracted a large number of emerging forces. The new players attempting to devote themselves to commercial spaceflight applications have brought unique business models that further promote the development of related industries. Meanwhile, the great progress in minisatellite technology has revealed the great value of large-scale LEO minisatellite constellations in a series of applications such as space-based global communication and remote sensing. The launch plans of large-scale LEO constellations are emerging explosively in an endless stream. Another big wave of LEO constellation development was seen in the 1990s. At that time, most of the LEO communication constellation concepts failed to enter the engineering stage. The only three constellations in operation were Iridium System, Globalstar, and ORBCOMM. However, they also struggled for a time, and all went through bankruptcy protection and reorganization. The following are the typical LEO satellite constellations currently deployed [4]:

(1) Iridium NEXT is the second generation of the Iridium system. A global space-based network of LEO satellites connected by inter-satellite links provides high-quality voice and data communications covering the entire Earth's surface including the oceans and polar regions. The satellites provide high-speed services of 1.5 Mb/s in the L band and 8 Mb/s in the Ka band. In addition to the communication payload, each satellite also carries the navigation augmentation payload and Automatic Dependent Surveillance-Broadcast (ADS-B) payload, which is used for aircraft flight monitoring and aircraft position tracking and reporting. On January 11, 2019, the eighth batch of "Iridium NEXT" satellites were successfully launched into LEO aboard a SpaceX rocket, marking the successful deployment of a new generation of constellation made up of 75 satellites.
(2) Globalstar communication constellation is a LEO-satellite mobile communication system operated by Globalstar, providing satellite voice and data movement services to enterprises, governments and individuals around the world. The 48 satellites in the space segment are distributed in 8 orbital planes, and can achieve full coverage between 70° north latitude and 70° south latitude. The ground segment is mainly composed of gateway stations, satellite operation control center, ground operation control center and Globalstar data network. Each gateway station communicates with three satellites simultaneously and synthesizes signals from different satellite data streams.
(3) Orbcomm IoT constellation is a global non-real-time low-rate data transmission system developed by Orbcomm, and is the first commercial satellite network in the world to provide M2M Internet of Things services. Currently, the constellation has 40 satellites in orbit, including 24 first-generation Orbcomm satellites (OG1) and 16 s-generation satellites (OG2). The Orbcomm will launch more satellites in the future to further reduce the network delay and enhance

the service level. The ground segment is mainly composed of ground-based gateways, control centers, and mobile and fixed satellite communication unit user terminals all over the world, currently including 16 ground-based gateway stations, 3 ground-based AIS data receiving stations and 3 regional gateway control centers.

(4) "Dove" Earth-observation constellation is the world's largest commercial Earth-imaging satellite constellation built by Planet Company, USA. In the constellation, 195 3U Cube Sats have been deployed. The satellite image accuracy is 3 m, which can distinguish trees, cars and ships, but is not high enough to identify pedestrians. Each day, the image data of about 550 gigabytes to 777 GB is down-linked and uploaded to the "Amazon Web Services" cloud computing platform. The Planet has acquired Google's TerraBella, a satellite image company, to further improve the image resolution to an expected accuracy of 1 m. The ground station network, including 11 global ground stations, has the advantages of low failure rate, serving multiple orbital planes, and short uplink and downlink transmission time.

The large-scale LEO constellation technology is making a comeback due to the rapid development of micro-satellites with new technology, fast response, flexible application and low cost and the great progress in space launch technologies, new manufacturing technologies, new imaging probe and image processing technologies, and IT technologies (represented by mobile communication technology and big data technology) made in the past decade. Multiple space powers and commercial companies in the world have put forward the launch and deployment programs for extremely large micro-satellite constellations. The typical new-generation LEO Internet constellations are listed below:

(1) OneWeb constellation, consisting of 882 (648 in orbit and 234 backup) satellites, is a LEO satellite constellation providing global coverage of Internet broadband services to individual consumers around the world. On June 22, 2017, this constellation was approved to enter the US market as the first new-generation non-geosynchronous orbit (NGSO) constellation program. In February 2019, the first six satellites were successfully launched. In the future, more satellites are expected to be launched.

    Each of the OneWeb satellites works in the Ku band, and can produce a throughput of 6 Gb/s. They adopt the "progressive pitch" technology to avoid interfering with Ku-band satellites in GEO. The ground-based user terminal antenna is a 36 cm x 16 cm phased array antenna and provides an Internet access speed of 50 Mb/s. In order to complete the networking work as soon as possible, the OneWeb will modularize the system components and use a large number of automation devices in the production line for mass production. According to this production model, 16 satellites will be produced per week to achieve an annual output of 648 satellites. The cost of satellite production can also be reduced.

(2) Starlink, the SpaceX's satellite Internet program, was approved by the U.S. Federal Communications Commission (FCC) to enter the American market in

March 2018. The constellation program plans to deploy a total of about 12,000 satellites, including 4425 MEO satellites at the altitudes of 1100–1300 km and 7518 LEO satellites at the altitudes of no more than 346 km. The test satellites (Tintin A and B) launched in early March 2018 are just two LEO satellites. According to SpaceX's estimation, the 12,000 satellites, whose deployment will be finally completed in 2025, are expected to deliver the broadband service of at least 1 Gb/s to Earth-based users and the ultra-fast broadband networks up to 23 Gb/s.

On May 24, 2019, 60 test satellites of the Starlink constellation were successfully launched, and were expected to collaborate with the two test satellites launched in February 2018 to further test the spaceborne antennas and electric propulsion systems. The SpaceX planned to launch two to six satellites in 2019 so as to officially start the constellation deployment.

(3) The Telesat Company in Canada also has a constellation program, which includes 117 LEO satellites. On November 3, 2017, the FCC approved Telesat's constellation to enter the American market for operation. The program is the second program to receive such approval following the OneWeb Constellation. On January 12, 2018, Telesat's first LEO high-throughput test satellite "LEO-Vantage1" was launched into orbit aboard the Indian Polar Satellite Launch Vehicle (PSLV). In 2021, the entire constellation began to operate with a terabyte-level communication capacity.

(4) In May 2018, the Russian Federal Space Agency (ROSCOSMOS) launched the Ether Global Multifunctional Communications Satellite System. The system consists of 288 satellites orbiting at an altitude of 870 km, providing mobile communication and Internet access services to users throughout Russia and around the world. It is the first LEO-satellite Internet constellation program proposed by Russia. According to the plan, the system will be connected with Russia's new-generation Gonets communication satellite system and ray-data relay satellite system, and is scheduled to be deployed by 2025. However, the program is making slow progress due to financial and technical problems.

The explosive development of large-scale LEO constellations has far exceeded the original expectations of the international community, while approaching the capacity limit of LEO space and bringing a series of severe challenges to the space environment. The ongoing deterioration of space debris environment has already threatened the LEO safety, and is still far from being contained. According to NASA's estimation, the LEO debris density, if not controlled effectively, will reach a critical level in 70 years, leading to a chained debris impact effect (Kessler Syndrome) that will render LEO space completely unusable. The launch and deployment of large-scale micro-satellite constellations will greatly aggravate this trend [5].

Firstly, because each large-scale constellation has a large number of satellites, its passing through an area of space object concentration in the launch process will greatly increase the number of space objects in the deployed orbit and the probabilities of collisions between space objects and between itself and debris. On June 28, 2019, three of the 60 micro-satellites launched by SpaceX lost track, two actively fell down, and 5% became debris. On September 2, 2019, the ESA Earth-observation satellite "Aeolus" took a maneuver to avoid a collision with the 60 micro-satellites launched by SpaceX. Secondly, due to the lower reliability compared with traditional satellites, the micro-satellites also face three problems: a high rate of injection failure, a high rate of operation failure, and a difficulty in guaranteeing the deorbit after reliable mission completion. Finally, the small target feature of micro-satellites makes it difficult for ground stations to effectively monitor, warn against and avoid them.

In the short term, the launch of large micro-satellite constellations will increase the number of LEO objects, thus threatening the safety of the existing space objects. Taking the OneWeb constellation consisting of 720 satellites at an orbital altitude of 1200 km as an example,[4,5] the number of dangerous rendezvouses between this constellation system and the cataloged space objects in three stages (launch deployment, on-orbit operation and post-mission deorbiting) is shown in Fig. 3.3. The number of dangerous rendezvouses in a single year during the OneWeb operation exceeded the number of dangerous rendezvouses of all the existing space objects in a whole year (more than 1000). In the five-year post-mission deorbiting stage, the OneWeb even had more than 10,000 dangerous rendezvouses. The number of dangerous rendezvouses can be significantly reduced by shortening the deorbiting time, which, however, has not been constrained by the existing debris mitigation guidelines. This is the effect of just a 720-satellite constellation, let alone the effect of a constellation consisting of tens of thousands of satellites.

In the long term, the large micro-satellite constellations themselves will also encounter debris impacts, which will increase the number of LEO impacts, generate more space debris, and aggravate the long-term deterioration of debris environment. The number, quality and area of satellites in a constellation, as well as the deployment altitude of the constellation, will have an impact on the collision frequency of the constellation in space environment, and the amount of new debris and future space debris. Among these factors, the number of satellites in the constellation is an important factor influencing the future space debris environment, the area of satellites mainly influences the number of future collisions, and the quality of satellites is an important factor influencing the amount of new debris. The deployment of the constellation at an altitude of 1100 km has the greatest impact on the space debris environment, while the deployment at an altitude of 500 km has the smallest impact. However, in order to achieve the mission objectives, the satellite orbit maintenance is required, bringing very high cost [6]. The measures for debris generation mitigation are needed to address the impact of large constellations on the space debris environment. Generally, the debris mitigation measures mainly include: passivation; limiting the objects released during normal operation; PMD and on-orbit collision avoidance. Among them, PMD is to transfer the spacecraft or rocket at the end of its mission with a future on-orbit life of more than 25 years from its original orbit

[4] Christian P. Provide an example of impact probability assessment for constellations above 1000 km. IADC–WG3–IT34–4, 2016.

[5] James Hyde, Eric Christiansen, Dana Lear. Impact probabilities for a constellation above 1000 km. 35th IADC–IT34–4, 2017.

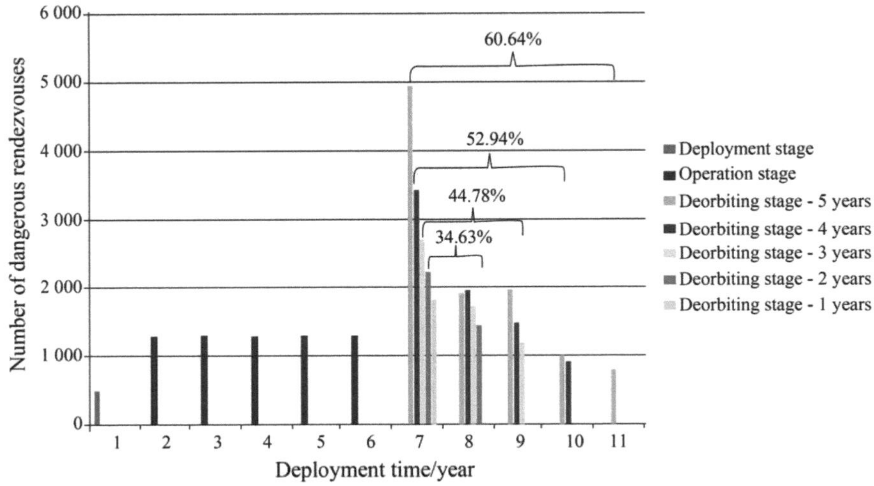


**Fig. 3.3** Number of dangerous rendezvouses of OneWeb constellation in different stages (see the colored illustration)

to a abandoned orbit, which allows the spacecraft or rocket to fall from sky naturally within 25 years, according to a certain success rate of disposal. The research shows that,[6] the launch of a typical large constellation (made up of 1080 200 kg satellites distributed in 20 orbital planes at an altitude of 1100 km and an inclination of 80°) with a lifetime of 50 years (2021–2071) will lead to a surge in the number of LEO objects and impacts in a short time. Only when the success rate of post-mission deorbit (PMD) of constellation satellites is at least 90%, the number of orbiting objects and the number of impacts will slowly return to the normal level after the decommissioning of the constellation. However, if the PMD is below 90%, the number of orbiting objects will keep growing, and the space environment will keep deteriorating irreversibly, as shown in Figs. 3.4 and 3.5. When the 90% PMD is achieved, the number of objects in LEO region and the peak number of impacts will decrease with the shortened deorbiting time, and the environmental recovery speed will increase accordingly, as shown in Figs. 3.6 and 3.7.

The existing space debris mitigation guidelines developed by the United Nations Committee on the Peaceful Uses of Outer Space (UNCOPUOS) and the Inter-Agency Space Debris Coordination Committee (IADC) are utterly incapable of containing the catastrophic impact of large-scale micro-satellite constellations on the space debris environment. To this end, the IADC submitted a statement on large-scale constellations to the UNCOPUOS General Assembly each year from 2015 to 2018, pointing out that "large-scale constellations will lead to serious deterioration of the space debris environment, posing a severe challenge to the existing mitigation guidelines."

[6] Holger K. Risk to space sustainability from large constellations of satellites, the 67th International Astronautical Congress, IAC–16–A6. 4.2, Guadalajara, Mexico, 2016.

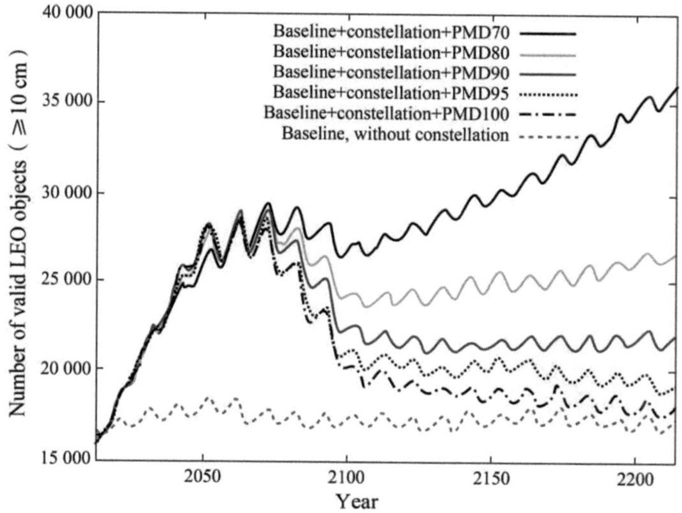


**Fig. 3.4** Effect of constellation on the amount of debris larger than 10 cm in LEO orbit at different PMD rates (see the colored illustration)

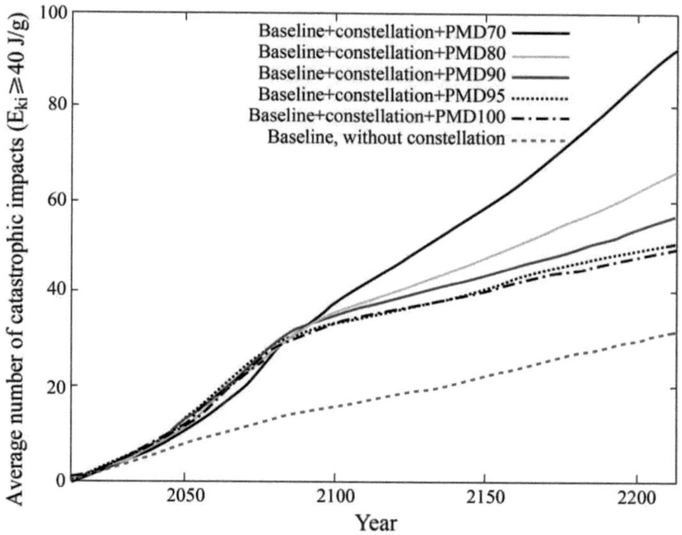


**Fig. 3.5** Effect of constellation on the number of LEO object impacts at different PMD rates (see the colored illustration)

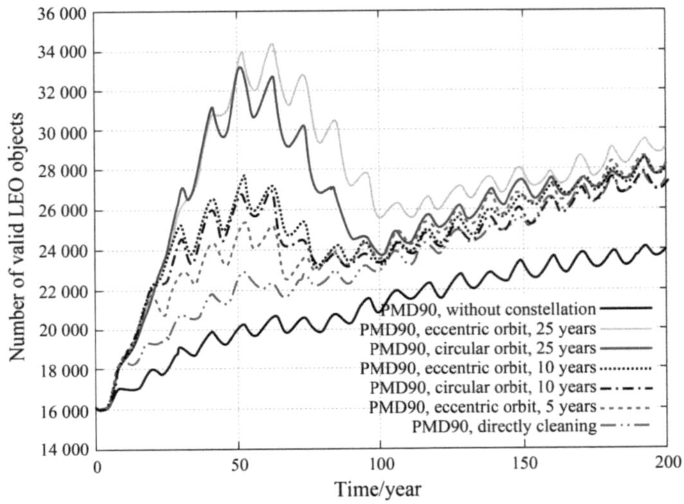


**Fig. 3.6** Effect of constellation on the amount of debris larger than 10 cm in LEO orbit at different PMD time (see the colored illustration)

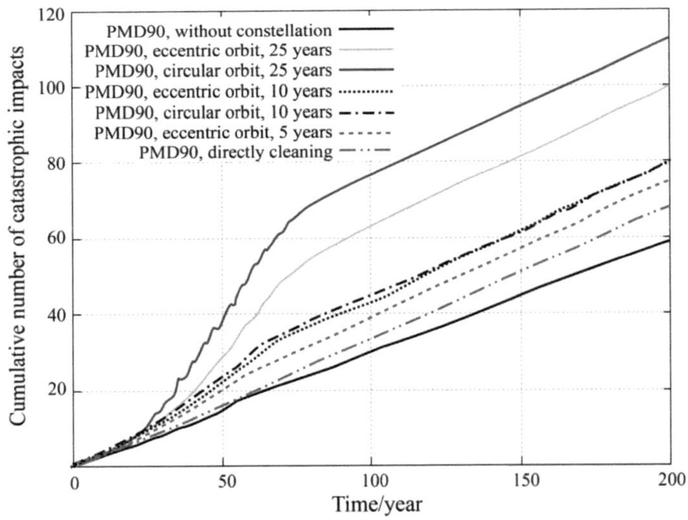


**Fig. 3.7** Effect of constellation on the number of LEO object impacts at different PMD time (see the colored illustration)

The IADC suggested that: (1) Large micro-satellite constellations should not overlap at orbital altitudes to avoid collisions with each other; (2) Any micro-satellite at an orbital altitude above 400 km should be equipped with a corner reflector, which could inform the monitoring system of the satellite's spatial position, so as to avoid collisions with other satellites; (3) The micro-satellites should be passivated and prepared for deorbiting rapidly and reliably after the end of their missions, and should be deorbited within 5 years after EOL: (4) Appropriate materials should be selected for the manufacture of micro-satellites to ensure the complete disintegration and burnout of satellites during re-entry into the atmosphere and no debris arrival at the ground to pose a significant risk to people or property on the ground.

## References

1. Liu, L. (2006). *Introduction to aerospace dynamics*. Nanjing University Press.
2. Xiao, Y. (1995). *Principles of spacecraft flight dynamics*. China Aerospace Press.
3. Hu, M., Fan, L., & Ren, Z. (2015). Research status and analysis of space traffic management. *Journal of Command and Control, 1*(3), 310–313.
4. Fang, F., & Wu, M. (2020). Development of global low-orbit satellite constellations. *Cruise Missiles, 425*(5), 95–99.
5. Zhao, Q., Hu, C., Chen, C., et al. Opportunities and challenges for large-scale LEO constellations. *Space Debris Research, 20*(1), 1–9.
6. Dan, S., & Jing, L. (2020). Analysis on the impact of large LEO constellation deployment on space debris environment. *Systems Engineering and Electronics, 42*(9), 2041–2051.

# Chapter 4
# Early Warning System for Space Debris

## 4.1 Overview of Space Debris Collision Warning

Avoiding space debris is almost a "routine maneuver" for space powers to ensure the safety of their spacecrafts and space stations. In recent years, the International Space Station (ISS) has adjusted its orbit and altitude several times, even twice in three weeks in the extreme case, to avoid space debris. According to statistics, the ISS made 14 maneuvers to avoid the impact threat of space objects from 2008 to 2014 (see Table 4.1), including 1 maneuver in 2008, 2 maneuvers in 2009, 1 maneuver in 2010, 3 maneuvers in 2011, 3 maneuvers in 2012 and 4 maneuvers in 2014. The increasing of avoidance maneuver frequency from 1 maneuver in 2008 to 4 maneuvers in 2014 also reflects the deterioration of spatial operating environment to some extent [1].

Even so, due to the limitation of observation resources and orbital dynamics model accuracy, the confidence level of collision warning is not very high at present. With the continuous growth of space debris, the risk of impact threat to the orbiting spacecrafts is increasing. On October 1–10, 2015, the United States reported as many as 21 collision warnings of China's on-orbit LEO satellites, all of which were later confirmed to be false. The warning-informed spacecraft avoidance will consume a large amount of the limited spaceborne energy, thus greatly shortening the useful life of the spacecrafts. Moreover, the frequent avoidance will disable the normal application of spacecrafts, so that their utilization efficiency will be greatly reduced. Therefore, reducing the false alarms and improving the confidence level of early warning is an important prerequisite for the engineering implementation of spacecraft collision warning and avoidance.

The direct reason for high false alarm rate is that the orbital positions of a spacecraft and a space object dangerous to the spacecraft are not accurate enough. If the 24 h prediction accuracy of orbital positions of the two objects is better than meter level, the 24 h collision warning for a meter-sized spacecraft can achieve a confidence level of 100%. However, the current situation is that the quasi-real-time orbit determination (OD) accuracy for spacecrafts with cooperative measurement can reach the



J. Zhang et al., *Space Debris Research*, https://doi.org/10.1007/978-981-96-9570-6_4

**Table 4.1** Statistics of ISS collision avoidance events

| Time of avoidance | Avoided object | Adopted measure |
|---|---|---|
| Aug 27, 2008 | Debris of Russian "Cosmos-2421" satellite | Collision avoidance maneuver |
| Mar 22, 2009 | Upper stage of China's carrier rocket | Collision avoidance maneuver |
| Jul 18, 2009 | Upper stage of Russian carrier rocket | Collision avoidance maneuver |
| Oct 26, 2010 | Debris of American UARS satellite | Collision avoidance maneuver |
| Apr 2, 2011 | Debris of Russian "Cosmos-2251" satellite | Collision avoidance maneuver |
| Jun 28, 2011 | Debris of disintegrated "Proton" rocket | The astronauts entered the Soyuz spacecraft for sheltering |
| Sep 29, 2011 | Debris of Russian "Tsyklon" rocket body | Collision avoidance maneuver |
| Jan 13, 2012 | Debris of "Iridium 33" satellite | Collision avoidance maneuver |
| Jan 28, 2012 | Debris of FY-1C satellite | Collision avoidance maneuver |
| Mar 24, 2012 | Debris of Russian "Cosmos 2251" satellite | The astronauts entered the Soyuz spacecraft for sheltering |
| Mar 16, 2014 | Debris of Russian "Meteor 2–5" satellite | Collision avoidance maneuver |
| Apr 3, 2014 | ESA "Ariane" rocket body | Collision avoidance maneuver |
| Oct 27, 2014 | Debris of Russian "Cosmos 2251" satellite | Collision avoidance maneuver |
| Nov 12, 2014 | Debris from Chinese RS-12 satellite and ESA "ATV_5" cargo spacecraft | Collision avoidance maneuver |

meter level or even centimeter level, while the quasi-real-time OD accuracy limit for dangerous space objects based on non-cooperative measurement can only reach the level of 10 m. A meaningful alarm generally shall be given at least 24 h before the impact. Therefore, the completion of quasi-real-time OD shall be followed by 24 h orbit position prediction through orbit dynamics extrapolation to achieve the collision warning calculation. Considering the influence of the accuracy of orbit dynamics model (especially the error of atmospheric density model), the accuracy of 24 h orbit position prediction can hardly reach the level of 100 m even if there is no error in quasi-real-time OD. We can imagine how low the probability of collision between two meter-sized (or smaller) objects within two 100 m-diameter error balls is. However, due to the limitation of global space detection resources, the existing international space-object cataloging system can only keep the quasi-real-time OD accuracy of all space objects at the kilometer level and the accuracy of 24 h orbit position extrapolation at the level of 10 km, so the collision warning based on the ordinary space-object cataloging system is basically untrustworthy.

Considering the current situation, we define two types of collision warning in the collision warning strategy: colorless warning and colored warning. Colorless warning is a collision alarm that is given relying on the accuracy of the cataloging system (i.e. without any additional tracking measurement) for conventional space object monitoring without special accuracy requirements. This warning can be dependent or independent on the device measurement information, but relying on the two-line elements (TLE) published on the Internet for collision warning. With too high false alarm rate and basically zero confidence level, this warning can not be used as a basis for spacecraft avoidance, but can be used to screen out dangerous events and objects and guide detection resources in targeted object observation. Colorless warning can be generally understood as the warning without additional observation cost under the conventional cataloging system, while colored warning is carried out after dangerous objects are picked out by colorless warning. Due to the need for guiding the avoidance, the colored warning has strong real-time performance and a high requirement for orbit prediction accuracy. Therefore, dangerous objects must be intensively observed. By stage and orbit accuracy requirement, colored warning is usually divided into yellow warning and red warning. Red warning is the premise of spacecraft avoidance, while yellow warning is the basis of red warning. The combination of the two types of warning is essential to further reduce false alarms of non-dangerous situations, focus more detection resources on dangerous objects, and improve the confidence level of red warning. In the stage of colored warning (especially red warning), all-time and all-weather stable detection equipment must be used to ensure the stability of data acquisition; a precision OD system must be used for OD and prediction calculation to ensure the high confidence level of relative position of a dangerous rendezvous point.

The establishment of a collision warning system with no missing alarm, low false alarm rate and high confidence level is based on not only a huge detection and measurement network but also accurate OD and prediction technology, while requiring a reasonable scheduling strategy for detection and computing resources and more importantly, an integrated, efficient and linked command and decision system. At first, the cataloging system continuously screens out a large amount of colorless warning information with possible collision risks to ensure that alarms are not missed within the range of space surveillance capabilities. Then, by using a reasonable resource scheduling strategy and focused observation, the yellow warning information still with collision risks within a reasonable range is screened out, and the vast majority of false alarms are eliminated. Finally, by using a precision OD prediction system, false alarms are minimized and important dangerous red events are identified to guide a spacecraft in collision avoidance. This is the thread of the whole collision warning effort.

## 4.2 Calculation Method of Space Debris Collision Warning

The collision warning methods for the orbiting spacecrafts mainly include the minimum distance method and the method based on impact probability. The minimum distance method is an average method that was first used by NASA. When the target entered the 5 km × 25 km × 5 km impact box in the RTN (radial, transverse and normal of the orbital plane) directions of the spacecraft, the focused monitoring of the rendezvous target should be carried out and the latest orbit parameters shall be used for calculation. When the target entered the 2 km × 5 km × 2 km region in the RTN directions of the spacecraft, the actual situation of the spacecraft should be considered to decide whether to avoid the target or not. However, the minimum distance method did not consider the location uncertainty and geometric size of the rendezvous target, which led to a high false alarm rate. Later, the relevant scholars proposed the impact probability method for early warning. In fact, both methods have limitations. Only through the combination of the two methods and the integrated implementation of dangerous objects screening, collision risk warning and avoidance strategy development in accordance with the object measurement and OD and the characteristics of different forecast time lengths, can a high degree of unity be achieved between theory and engineering implementation [1].

### *4.2.1 Dangerous Objects Screening*

So far, the total number of measurable space objects in orbit is about 16,000. If so many objects are tracked and their exact orbits are calculated one by one, the detection resources will be exhausted and the computing load will be too huge to imagine. To reduce the workload and improve the computational efficiency, the approaching process must be analyzed and dangerous objects must be screened out. In other words, before high-precision orbit prediction, the objects that are impossible to cross the orbit of the spacecraft of interest should be quickly excluded from a large number of on-orbit objects. Then the objects that may collide with the spacecraft or have a distance less than a certain value from the spacecraft should be screened out for further orbit prediction and error analysis.

(1) Use the altitudes of perigee and apogee to screen out dangerous objects

In order to identify in the possible shortest time the space objects that are likely to collide with the orbiting spacecraft, a large number of space objects shall be initially screened. The first criterion of screening is the perigee/apogee altitude. The space objects whose perigee altitude is higher than the apogee altitude of the spacecraft or whose apogee altitude is lower than the perigee altitude of the spacecraft are screened out, leaving the objects that may pose a threat to the spacecraft for further analysis. The changes of perigee and apogee altitudes mainly include the changes in the short term and the changes in the long term.

(2) Use the geocentric distances of intersections for screening

The second criterion of dangerous objects screening is the geocentric distance of intersection point. For two orbiting objects, their orbital planes will theoretically intersect each other at two points as long as their orbital inclinations or RAANs are not equal. However, a collision between the two space objects is likely to happen only at their intersections, so the geocentric distances of their intersections are also an important reference for coarse screening. In this way, dangerous objects are screened out according to the geocentric distances of the intersections, leaving the objects that may pose a threat to the orbiting spacecraft for further analysis.

(3) Use the minimum distance between two ellipses for screening

The third criterion of dangerous objects screening is the minimum distance between the elliptical orbits of two objects. The dangerous objects are screened out according to the minimum distance between two ellipses, leaving the objects that may pose a threat to the spacecraft for further analysis.

This screening process shall consider the geometric position relation of two ellipses in space. Two elliptic orbits must have two closest distances $d_1$ and $d_2$ at two points, which are respectively near the two intersection lines of the two elliptic orbital planes. If the minimum distance between the two ellipses is greater than the distance threshold, the collision between the two objects will be impossible. In all the presented three calculation methods, detailed orbit calculation is not needed, but four slow-varying parameters of the six orbital parameters, namely $a$, $e$, $i$ and $\Omega$, shall be used for qualitative judgment. In this way, the warning-related computational load can be reduced by one order of magnitude with the minimum computational effort and the fastest speed.

### 4.2.2 Minimum-Distance Calculation Method

The minimum distance and related parameters of two space objects in the process of approaching are very important factors for judging whether there is a risk of collision between them. If the orbit prediction error for the two objects is not considered, the minimum distance will be a decisive factor for judging the impact probability. The parameters involved in the minimum-distance calculation method include the minimum distance between two objects, the distance component in UNW (or RTN) direction, the approach velocity and the approach angle.

1. **Minimum Distance**

For the calculation and analysis of the relative minimum distance between two space objects, orbit prediction accuracy is the main factor affecting the confidence of the calculation results. Therefore, for the objects involved, a high-precision dynamical model must be adopted to predict their orbits. The analysis algorithm is mainly divided into the following three parts:

(1) Prediction of space object orbit

The dynamical model for object orbit prediction is:

$$\ddot{\boldsymbol{r}} = \boldsymbol{a}_0 + \boldsymbol{a}_e$$

$$\boldsymbol{a}_0 = -\frac{\mu}{r^3}\boldsymbol{r}$$
$$\boldsymbol{a}_\varepsilon = \sum_{i=1}^{6} \boldsymbol{a}_i \tag{4.1}$$

where $\mu$ is the gravitational constant of the Earth, $a_0$ is the central gravitational acceleration of the particle Earth, and $a_\varepsilon$ is the total perturbation acceleration. The perturbation models involved include Earth oblateness perturbation, third-body perturbation, atmospheric drag perturbation, solar radiation pressure perturbation, solid tide, sea tide perturbation and relativistic effect.

(2) Calculation of the minimum distance

If the position vector of the spacecraft is $r_1 = \left(x_1, y_1, z_1\right)^{\mathrm{T}}$ and the position vector of the space object is $r_2 = \left(x_2, y_2, z_2\right)^{\mathrm{T}}$, the relative distance between the two objects will be:

$$D = \sqrt{(x_1 - x_2)^2 + (y_1 - y_2)^2 + (z_1 - z_2)^2} \tag{4.2}$$

The relative distances between the spacecraft and the space object at all epochs can be calculated. From the whole set of relative distance values, the minimum distance can be found by interpolation.

(3) Relation between relative velocity and relative distance at the minimum distance

In the J2000 inertial coordinate system, the position vectors of the spacecraft and the space object at the minimum distance are assumed to be $r_1$ and $r_2$, and their velocity vectors are assumed to be $v_1$ and $v_2$ respectively. Then the relative distance vector between them is

$$\Delta \boldsymbol{r} = \boldsymbol{r}_1 - \boldsymbol{r}_2 \tag{4.3}$$

The approaching velocity between them is

$$\Delta \boldsymbol{v} = \boldsymbol{v}_1 - \boldsymbol{v}_2 \tag{4.4}$$

Since the distance value is the minimum at this time, we can obtain

$$\frac{\mathrm{d}}{\mathrm{d}t}|\Delta r|^2 = 2(\boldsymbol{r}_1 - \boldsymbol{r}_2)\frac{\mathrm{d}(\boldsymbol{r}_1 - \boldsymbol{r}_2)}{\mathrm{d}t} = 2(\boldsymbol{r}_1 - \boldsymbol{r}_2)(\boldsymbol{v}_1 - \boldsymbol{v}_2) = \Delta \boldsymbol{r} \cdot \Delta \boldsymbol{v} = 0 \tag{4.5}$$

The dot product of the two vectors is zero, which means that when the distance between the two objects is the minimum, their relative position vector and relative velocity vector are perpendicular to each other. As the feature of two approaching objects, this equation can be used as a correction formula for iteratively calculating the closest distance between the two objects so as to greatly improve the efficiency and accuracy of early warning calculation.

2. **Calculation of Distances in Three Directions**

In addition to the closest distance, the distance components in three directions under the UNW (or RTN) coordinate system should be used to more intuitively describe two objects approaching to each other.

In the UNW or RTN coordinate system with spacecraft as the center of mass, the components of the closest distance in the $x$, $y$ and $z$ directions are

$$\begin{aligned} \Delta \boldsymbol{r}_{\mathrm{UNW}} &= \boldsymbol{M}_{\mathrm{UNW}} \cdot \Delta \boldsymbol{r} \\ \Delta \boldsymbol{r}_{\mathrm{RTN}} &= \boldsymbol{M}_{\mathrm{RTN}} \cdot \Delta \boldsymbol{r} \end{aligned} \tag{4.6}$$

where $M_{\mathrm{UNW}}$ and $M_{\mathrm{RTN}}$ are the coordinate transformation matrices from J2000 inertial coordinate system to UNW and RTN coordinate systems, respectively.

3. **Accuracy Evaluation for Distance Components in Three Directions**

Among the collision warning thresholds, the distance threshold is mainly related to orbit prediction error, which, in turn, is directly related to orbit determination error and prediction model. The orbit determination error mainly depends on the accuracy of orbit dynamics model and the quality and quantity of the tracking data of detection equipment. For an object in a nearly circular orbit, the following basic conclusions are drawn:

(1) The prediction errors in the radial and normal directions of the orbit are much smaller than the error in the track direction and are relatively stable.
(2) The errors of orbit position prediction are mainly concentrated along the track direction, and diverge rapidly with the square of time. The lower the orbital altitude, the more serious the divergence of errors. The propagation of orbit prediction errors along the track direction does not obey the zero-mean distribution, but skews to one side of the real orbit, thus resulting in a certain systematic deviation.
(3) The radial orbit position prediction errors also diverge with the square of time (more slowly than those along the track direction), and produce a certain model system error. The lower the orbital altitude, the more serious the divergence.
(4) The normal orbit position errors are relatively small, mostly at the level of tens of meters, and generally obey the zero-mean distribution under good observation data constraints.

4. **Calculation of Approach Angle and Approach Velocity**

In J2000 inertial coordinate system, the approach angles and approach velocities of the spacecraft and space object can be calculated by the following equation:

$$\theta = \arccos \frac{\boldsymbol{v}_1 \cdot \boldsymbol{v}_2}{|\boldsymbol{v}_1| \cdot |\boldsymbol{v}_2|}$$

$$\Delta v = |\boldsymbol{v}_1 - \boldsymbol{v}_2| \tag{4.7}$$

## 4.2.3 Method of Impact Probability Calculation

The calculation of impact probability is one of the bases for collision risk analysis. The trajectory safety analysis method based on impact probability uses impact probability as an indicator to describe the collision risk level. Impact probability is defined as the probability of collision between two space objects with position prediction errors. The calculation of impact probability needs the information about the position, velocity and position covariance matrix of the spacecraft and space object at the moment of encounter.

1. **Impact Probability**

The impact probability between two space objects is the probability that the center distance of the two objects is less than the sum of the equivalent radii, and can be expressed as $P_e = P(\rho < R)$. The distance between the two objects is $\rho = |\rho| = \left| r_1 - r_2 \right|$, where r1 and r2 are respectively actual position vectors of the two objects and can be respectively expressed as distribution center vectors of the two objects plus random error vectors, namely $r_1 = r_{1o} + e_1$, $r_2 = r_{2o} + e_2$.

According to Sect. 4.2.2, when the distance between the two objects is the closest, they are in a plane perpendicular to the relative velocity vector, which is defined as the encounter plane. In this way, the position uncertainties of the two objects can be projected onto the encounter plane, thus transforming the three-dimensional problem into a two-dimensional problem.

On the encounter plane, the impact probability density function $f(x, z)$ and impact probability $P_c$ can be calculated by the following equations:

The two-dimensional normal distribution probability density function is

$$f(x, z) = \frac{1}{2\pi\sigma_x'\sigma_z'} \exp\left\{-\frac{1}{2}\left[\frac{(x-\mu_x)^2}{\sigma_x'^2} + \frac{(z-\mu_z)^2}{\sigma_z'^2}\right]\right\} \tag{4.8}$$

where $\sigma'_x$ and $\sigma'_z$ are the components of the joint error in the encounter-plane coordinate system; $\mu_x$ and $\mu_z$ are the components of the relative distance between the two objects in the encounter-plane coordinate system.

The impact probability can be expressed as the integral of the two-dimensional normal distribution probability density function in the circle region:

$$P_c = \iint\limits_{x^2+z^2 \le R^2} f(x, z)\mathrm{d}x\mathrm{d}z \tag{4.9}$$

The main factors that affect the calculation result of impact probability include the size, relative position, relative velocity and position error of the two space objects.

2. **Maximum Impact Probability**

The theoretical and practical analysis shows that given a certain rendezvous position, velocity, geometric condition and object size, the impact probability will increase with the uncertainty of position error, and then reach its maximum under a certain position error, and then decrease with the increase of position error. In practical engineering applications, the calculation of the maximum impact probability or the impact probability in the worst case is very important, because the actual position error covariance matrices of the spacecraft and space object are usually unknown, or only the shapes (rather than the specific parameters) of their error ellipsoids are known. The maximum impact probability can also be used for the pre-screening of space objects. When the maximum impact probability is less than a certain preset threshold, the space object is considered posing no threat to the spacecraft.

Let $\mu_r = \sqrt{\mu_x^2 + \mu_z^2}$. The variable with a dimension of 1 is defined as:

$$v = \frac{\mu_r^2}{2\sigma^2}, u = \frac{R^2}{2\sigma^2} \tag{4.10}$$

The impact probability can be converted into an infinite series, whose first term is assumed to be the approximation of $P_c$:

$$P_c = e^{-r}\left(1 - e^{-\mu}\right) \tag{4.11}$$

The maximum value of $P_c$ can be obtained by letting $\frac{\partial P_c}{\partial \sigma} = 0$, so

$$\begin{aligned} \frac{\partial P_c}{\partial \sigma} &= \frac{\partial P_c}{\partial v} \cdot \frac{\partial v}{\partial \sigma} + \frac{\partial P_c}{\partial u} \cdot \frac{\partial u}{\partial \sigma} \\ &= e^{-v}\left(1 - e^{-u}\right)\frac{\mu_r^2}{\sigma^3} - e^{-v}e^{-u}\frac{R^2}{\sigma^3} \\ &= 0 \end{aligned} \tag{4.12}$$

Then the Eq. (4.12) is solved to obtain

$$u = \ln\left(1 + \frac{R^2}{\mu_r^2}\right) \tag{4.13}$$

Therefore, when $\sigma = \sigma_D$, $P_c$ reaches its maximum $P_{c\max}$, i.e.

$$\sigma_D = \frac{R}{\sqrt{2\ln\left(1 + \frac{R^2}{\mu^{r1}}\right)}} \tag{4.14}$$

$$P_{cmax} = \frac{R^2}{R^2 + \mu_r^2}\left(\frac{\mu_r^2}{R^2 + \mu_r^2}\right)^{\frac{\mu_r^2}{R^2}}. \tag{4.15}$$

## 4.3 Debris Collision Warning Strategies

With the continuous improvement of the utilization rate of space resources, the number of space objects in space is increasing sharply, the orbit environment of spacecrafts is deteriorating, and the probability of collision between space objects is rising. An impact to any orbiting spacecraft will bring immeasurable losses to national economic construction and national defense security. To ensure the safety of space missions, early collision alarms must be given in time to guide spacecrafts in collision avoidance control. Meanwhile, to minimize the avoidance cost caused by spacecraft warning, a complete set of organization processes and norms should be developed, and an integrated collision warning and avoidance system should be established.

According to the accuracy of orbit prediction and the time required for avoidance disposal, the collision warning and avoidance process of an orbiting spacecraft is generally divided into four stages: dangerous objects screening (colorless warning), daily warning analysis (yellow warning), fine warning (red warning), and avoidance control.

1. **Dangerous Objects Screening**

Every day, the normal cataloged orbits are used to complete the 7-day (from that day) collision warning calculation for all the orbiting spacecrafts. If there is an approach event that enters the threshold of potential danger warning, a supplementary tracking plan shall be issued for the involved dangerous object, and the tracking and measurement data shall be accumulated to lay a foundation for further judging the risk level of the approach event.

2. **Daily Warning Analysis**

While maintaining the normal cataloging OD system without increasing the workload of the detection network, the collision warning analyzer shall screen out dangerous space objects with 7-day collision risks every day. The selection of a specific threshold in the minimum distance method is strongly related to the accuracy of detection equipment, the size of space object, the inclination and altitude of space object orbit, as well as the accuracy and release time of orbit calculation software model. In the stage of dangerous objects screening, simple-numeric cataloging orbit calculation method is preferred due to the combined requirements of calculation accuracy and efficiency. Simple numeric method, American TLE method, average element method and semi-numerical semi-analytical method can all be used in orbit calculation and collision warning calculation at this stage. However, it is recommended that only the numerical orbit calculation method with the most precise model be used in the orbit calculation of this stage and later stages. Generally, daily collision warning is referred to as yellow warning if the rendezvous time is no more than 3 days later as predicted by precise orbit calculation after dangerous objects are screened out and are intensively observed through "two ascends and two descends".

Given no alarm dismissal, the yellow warning can provide high confidence and greatly reduce the occurrence of false alarm events. The determination of an appropriate yellow warning threshold should reduce the possibility of false alarm to a level basically not affecting the service life and normal operation of the orbiting spacecraft while ensuring no alarm missed. Considering the current space object density, detection network capability and overall spacecraft performance, each spacecraft should be given false yellow warning for no more than 3–5 times per year.

3. **Fine Warning**

After yellow dangerous objects are identified and yellow warning events are issued, the detection network will focus its resources on the maximized tracking and observation of few yellow dangerous objects in the next 48 h to ensure that the highest-precision orbits can be obtained 24 h before dangerous rendezvous time. Within 48 h after a yellow warning is given, the detection network should precisely measure the orbit of the dangerous object in at least two ascends and two descends or preferably five ascends and five descends. The precision orbit determination based on the observation data of two ascends and two descends shows that the position error of 24 h orbit prediction is basically maintained at the level of hundreds of meters, which can barely support the early warning work with a certain level of confidence. The precision orbit determination based on the observation data of five ascends and five descends can control the position error of 24 h orbit prediction within 100 m and can greatly improve the confidence level in the collision warning for the meter-sized or even 10 m-sized spacecraft. Here, a warning event that cannot be excluded by the work of this stage is called red warning.

Most of false alarms in yellow warning can be effectively eliminated by using an appropriate radial threshold. The red warning given in this case has high confidence. In red warning, the selection of impact probability for a space object of normal size should ensure that at least one real collision may hide in no more than 10,000 dangerous red warning events. However, considering the serious damage caused by a collision, avoidance control is suggested for the spacecrafts with orbit control ability, especially manned spacecrafts or high-value spacecrafts, to ensure their absolute safety. The confidence in red warning will be somewhat reduced when the possible occurrence of a geomagnetic storm in the next 24–48 h is indicated by sunspot observation. In this case, the collision warning probability and maximum probability can be calculated to comprehensively determine whether or not to start the avoidance control.

The closer the time of red warning is to the dangerous collision point, the higher the orbit prediction accuracy and the confidence of the red alert given will be. However, it usually takes the spacecraft a long time to prepare for orbit control. Theoretically, the earlier control requires less fuel and produces a more obvious effect of avoidance isolation. Therefore, from the point of view of avoidance, the earlier control is better. Considering the 24 h controllability of atmospheric drag model error and the feasibility of intensive observation by detection network, our suggestion is that the red collision warning be given to a LEO spacecraft one day in advance.

4. **Avoidance Control**

The collision warning work enters the avoidance control stage when the dangerous rendezvous time is 24 h apart. In general, the avoidance control time should be 6 h earlier than the dangerous rendezvous time. At first, based on the theoretical post-control orbit of the spacecraft, the collision warning analysis and calculation should be made for 3 days after the spacecraft avoidance. Then the safety of the theoretical post-control orbit should be evaluated. If there is a yellow warning event, the theoretical post-control orbit should be recalculated and the above process should be repeated until no yellow warning event occurs.

The selection of orbit avoidance strategy is related to not only the ability and state of each direction control mechanism of the orbiting spacecraft, but also the level of orbital control preparation and decision-making of the spacecraft management department. In most cases, the follow-up mission of the spacecraft also needs to be taken into account. Therefore, the selection of orbit avoidance strategy is usually a complicated process. In whatever algorithm, the theoretical target orbit for avoidance control needs to be re-warned and re-checked in advance to ensure that no new collision risk will occur in the avoided orbit within a certain period of time.

## 4.4 Design of Debris Collision Warning System

In terms of collision warning system design, the Orbital Debris Model (ORDEM) of NASA Orbital Debris Program Office (ODPO) has been widely used around the world and the ORDEM3.0 has been released. In this section, the debris collision warning system [2] and collision risk assessment system[3] designed by Chinese researchers are mainly presented.

### *4.4.1 Debris Collision Warning System*

1. **Overall System Framework**

The main program framework of the system is shown in Fig. 4.1. The sub-modules of database management system and warning analysis system are organically combined into a complete system.

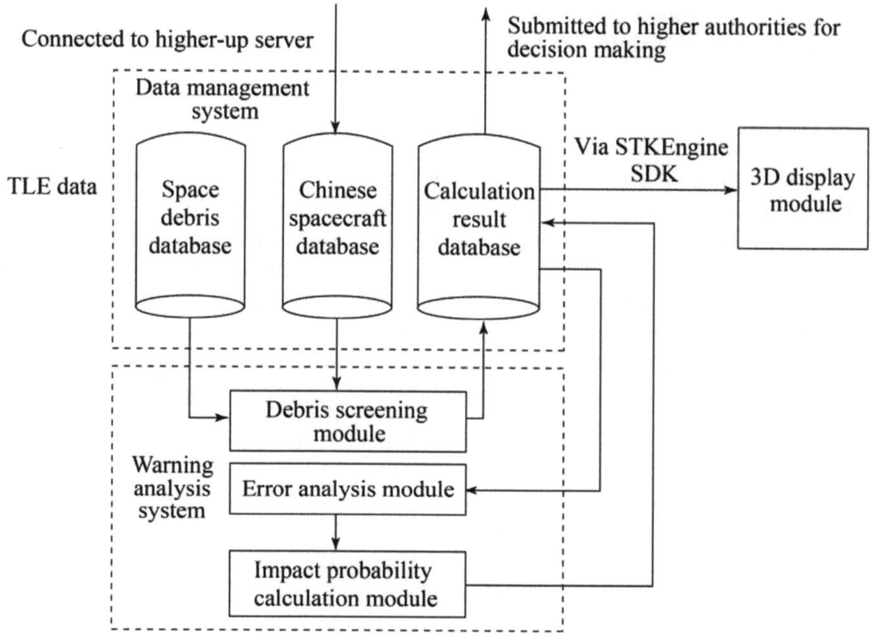


**Fig. 4.1** Main program framework of collision warning system

The space debris database and calculation result database in the database management system provide data services in the form of Oracle database. The Chinese spacecraft database mainly stores the osculating elements and initial orbit error information in the form of XML file coming from higher authorities. The space debris database stores the TLE data in TLE format from celestrak.com. The data in celestrak.com comes from U.S. Space Surveillance Network (SSN) and third-party orbiting spacecraft owners, and some of the data is updated more than once a day. Therefore, the update frequency of TLE data in the system is set as 3 times a day.

After a series of preliminary screening of space debris, the warning analysis system will accurately calculate the impact probabilities of the remaining space debris that may collide with Chinese spacecrafts. The information about the space debris with higher impact probabilities will be stored in the database and submitted to the higher authorities for decision-making.

2. **Debris Screening Module**

The amount of the existing cataloged space debris is large and will grow with the increase of human space activities. The direct calculation of impact probability of each debris piece will consume a lot of time so that the warning calculation process will become too slow to play the role of timely prediction. In this module, the epoch time method, perigee and apogee method and minimum distance method are mainly used to preliminarily screen the space debris, thus greatly improving the efficiency of the whole warning process. In addition, some intermediate results can be directly used for impact probability calculation in the next step. The flow chart of debris screening process is shown in Fig. 4.2.

3. **Error Analysis Module**

The error analysis module is mainly used to calculate the position error of the spacecraft and space debris when the spacecraft is closest to the space object. For Chinese spacecrafts, this error is calculated by using the initial covariance matrices periodically provided by the superior department. When calculating the position error at the time of closest approach, the covariance matrix closest to the time of closest approach should be selected. The HPOP Covariance instruction set in STK Engine SDK is used to set the corresponding parameters, which are then used to deduce and calculate the covariance matrix. The main instructions are as follows:

(1) The prediction of position covariance matrix is activated using the instruction "hpop*/Satellite/Satellitel Covariance On".
(2) The order of gravitational field model is set using the instruction "hpop*/Satellite/Satellitel Covariance Gravity".
(3) The atmospheric drag coefficient is set using the instruction "hpop*/Satellite/Satellitel Covariance ConsiderParam Drag".
(4) The solar radiation pressure coefficient is set using the instruction "hpop*/Satellite/Satelitel Covariance ConsiderParam SRP".
(5) The coordinate system for position error representation, namely J2000 inertial coordinate system, is set using the instruction "hpop*/Satellite/Satellitel Covariance Frame".
(6) The initial position error is set using the instruction "hpop*/Satellite/Satelitel Covariance PosVel".

Since space debris is a non-cooperative target and accurate observation data is unavailable, the variance matrix of its position error can hardly be accurately calculated. Therefore, in the error analysis module, the TLE historical data is used to fit the covariance evolution function, which, in turn, is used to calculate the covariance matrix at the time of closest approach.

4. **Impact Probability Calculation Module**

Based on the obtained positions, velocities and error covariance matrices of the spacecraft and space debris at the time of closest approach, the impact probability calculation module uses the Patera impact probability calculation method recommended by the "Crossing Data Message" (CDM) standard of the Consultative Committee for Space Data System (CCSDS) to calculate the impact probability, and then stores the results in the database. In the module, the relative velocity of the spacecraft and space debris is assumed to be very large and their rendezvous time is assumed to be very short. Therefore, the relative motion of the two objects is uniform linear motion within the rendezvous time without velocity uncertainty so that the position error ellipsoid remains unchanged during the rendezvous time.

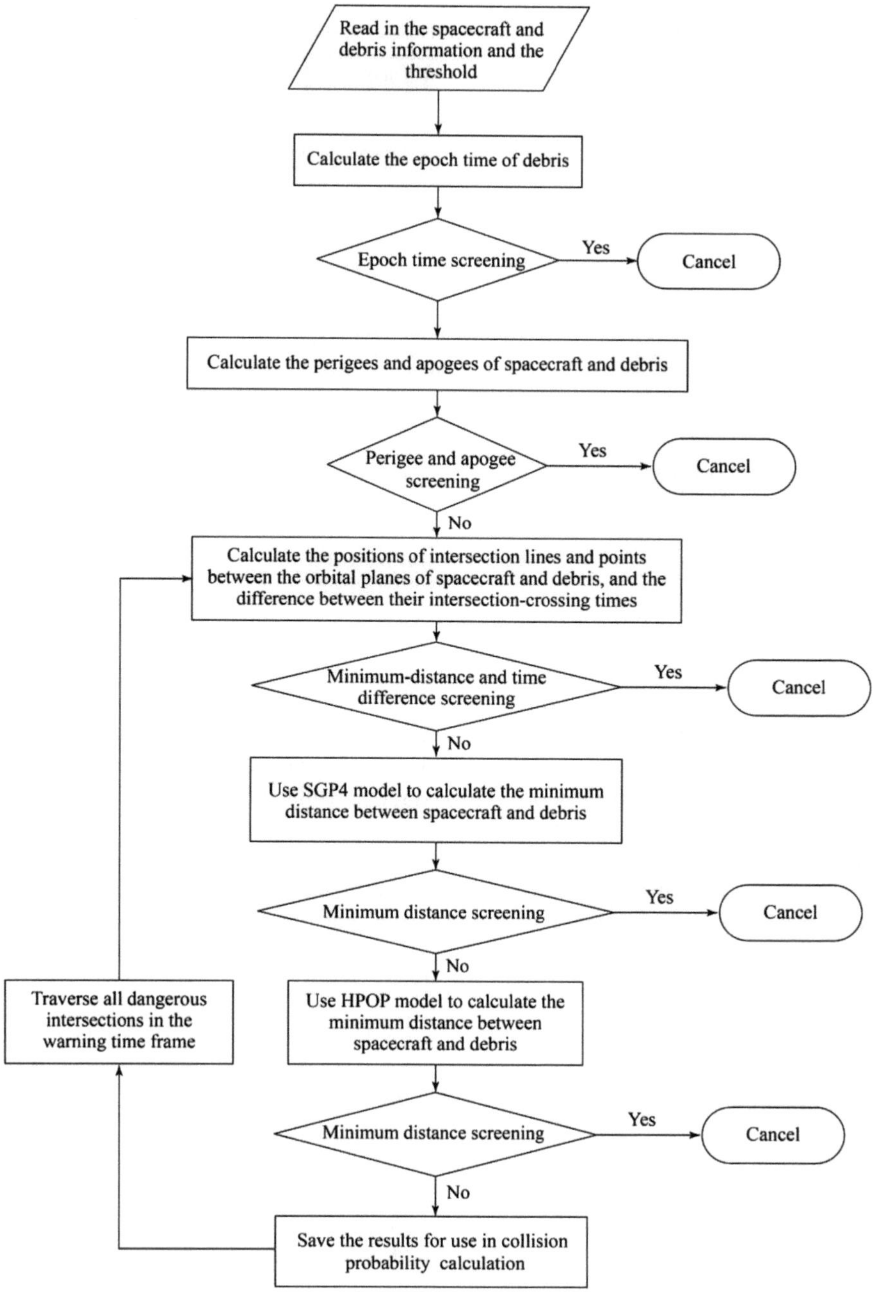


**Fig. 4.2** Flow chart of debris screening process

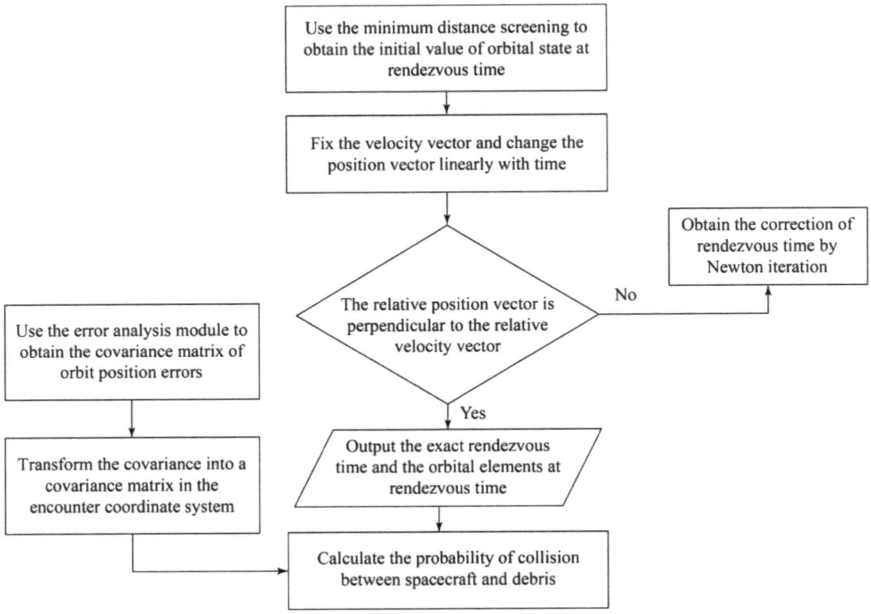


**Fig. 4.3** Impact probability calculation process

In the debris screening module, the step time is set as 0.1 s by the system to accelerate the calculation speed when using HPOP model to calculate the orbital state of spacecraft and space debris. However, the displacement of a LEO spacecraft can reach about 700 m within 0.1 s, so the orbital state calculated in this way is not accurate enough. In order to calculate the orbital state at the time of closest approach more accurately, the system uses the Newton iteration method for orbit state correction in accordance of the criterion that the relative position vector is perpendicular to the relative velocity vector at the time of closest approach. The impact probability calculation process is shown in Fig. 4.3.

### *4.4.2 Debris Collision Risk Assessment System*

1. **Overall System Framework**

The overall system framework is shown in Fig. 4.4. By controlling the interface program, the main program of the module of debris risk assessment system (MODRAS) organically combines the application program system, information management system and pre- and post-processing system into a complete system. The MODRAS consists of framework platform design, application program module design, and service software.

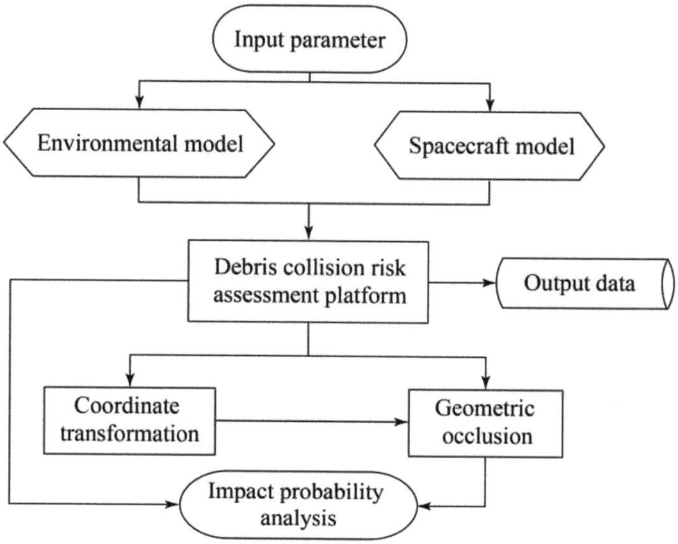


**Fig. 4.4** Overall framework of MODRAS

2. **Space Environment Module**

The space environment module provides M/OD direction for the geometric occlusion module and provides M/OD flux for the impact probability calculation module. The space environment module in MODRAS system includes three types of analysis codes: ORDEM2000, NASA91 and NASA SSP-30425. The space debris environment module "NASA91" and micrometeoroid module "NASA SSP-30425" are both given analytically and are relatively easy to deal with, while the space debris environment model "ORDEM2000" only outputs environmental data files (∗.dat). Therefore, the environmental data shall be read by interface program and then the data such as debris flux and direction shall be stored in a certain format.

ORDEM2000 is a LEO (200–2000 km) debris environment model outputting the debris fluxes with six different diameters: $\geq 10$ μm, $\geq 100$ μm, $\geq 1$ mm, $\geq 1$ cm, $\geq 10$ cm, $\geq 1$ m. To evaluate the spacecraft collision risk, the space particle data of any size is needed. According to the user manual of ORDEM2000, the flux of the debris of any size between 10 μm and 1 m can be obtained by cubic spline interpolation. The program design is shown in Fig. 4.5.

3. **Spacecraft Finite-Element Modeling Module**

Since each part of the spacecraft faces a different risk of encountering the M/OD, a finite element model shall be established for the spacecraft before its collision risk analysis. The finite element modeling (FEM) is realized by MODRAS through calling the FEM software "TrueGrid". The FEM program is shown in Fig. 4.6.

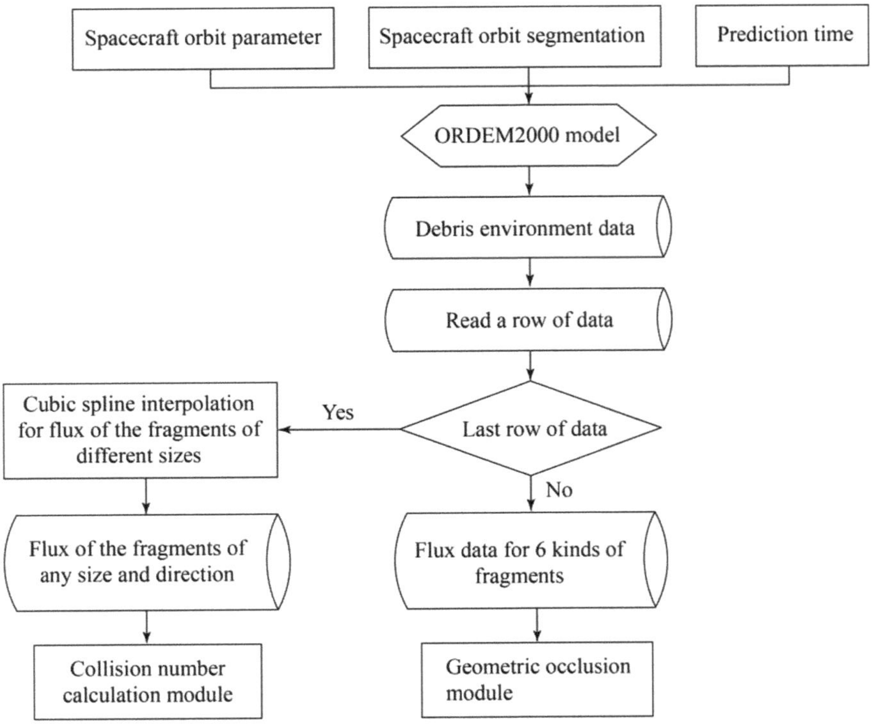


**Fig. 4.5** Program of ORDEM2000 debris environment module

4. **Geometric Occlusion Processing Module**

The correct handling of the spacecraft's geometric occlusion is a key step in collision risk assessment, as shown in Fig. 4.7. In a certain direction of debris inflow, only part of the spacecraft surface will be hit by the coming particles. How to determine the spacecraft surface area possibly hit by each particle is the core problem to be solved by geometric occlusion module. The basic principle of geometric occlusion processing is to divide the spacecraft into several segments, delete the occluded elements in each segment, and then merge the segments together and delete the occluded elements between them.

5. **Space Debris Collision Module**

The space debris collision module is an important part of the debris collision risk assessment system. The module outputs two types of assessment results, collision number and collision probability, which are the important data results of spacecraft collision risk assessment. The M/OD collision number and collision probability are obtained by taking the output results of the above-mentioned modules as inputs to the collision probability calculation formula. The program of this module is shown in Fig. 4.8.

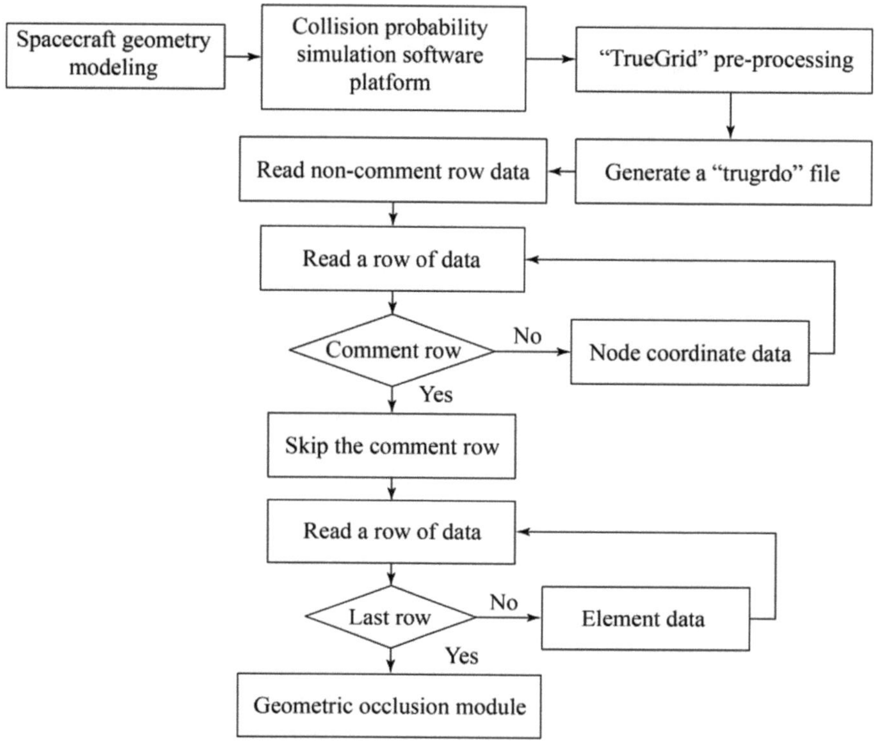


**Fig. 4.6** FEM program

## 4.5 Development Trend of Debris Collision Warning Research

With the constant progress in science and technology, the research of debris warning is being gradually deepened. The precise warning and comprehensive warning will be used in practice. Meanwhile, to better avoid the danger of debris collision, the future spacecrafts are expected to autonomously identify the threatening debris and actively avoid the debris collision [4].

1. **Precise Warning**

With the development of space observation technology, the new devices for debris observation are increasing. The phased array radar can increase the number of the fragments being observed to 5000–10,000 station circles per day. The "Space Fence" with powerful S-band radar proposed by the U.S. Air Force Electronic System Center can track a piece of debris as small as 1 in[1] above the southern Hemisphere. The

[1] 1 in ≈ 2.54 cm.

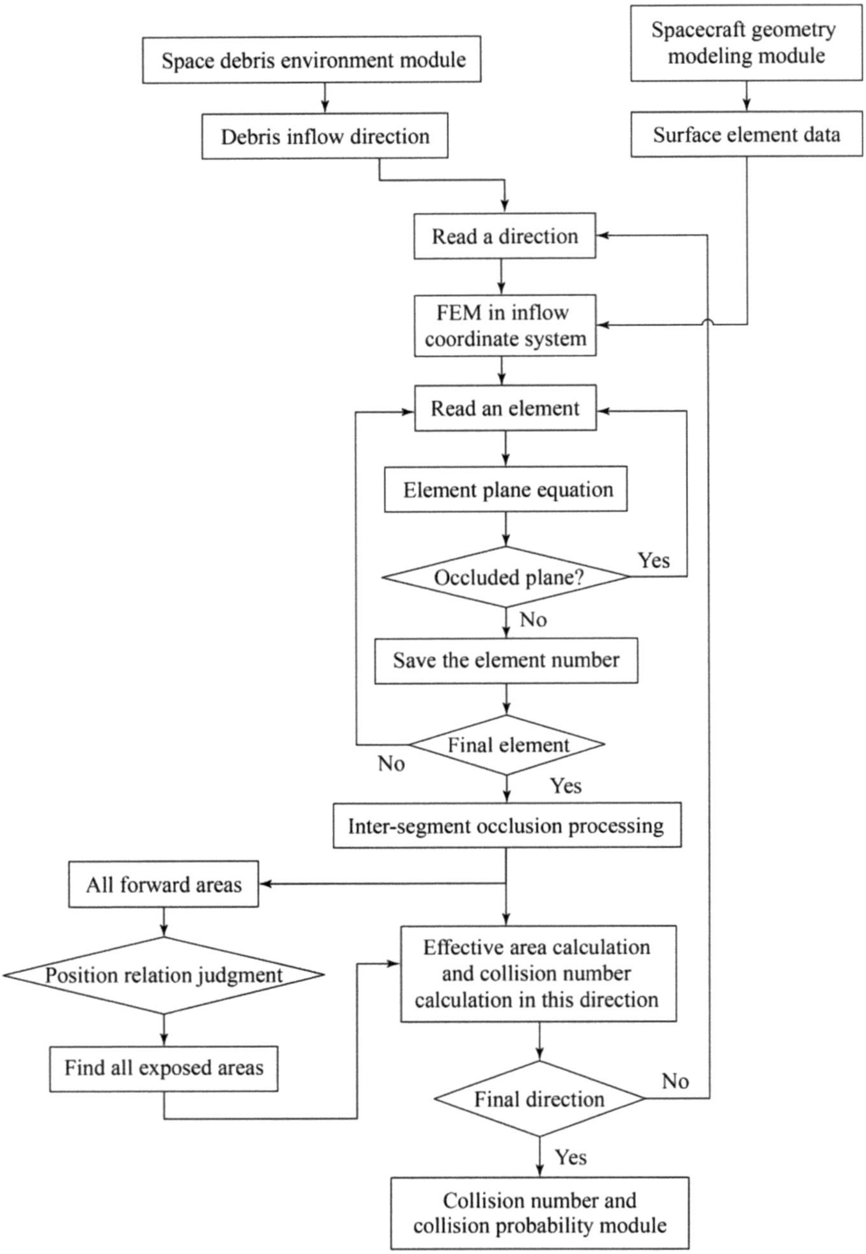


**Fig. 4.7** Program of geometric occlusion processing module

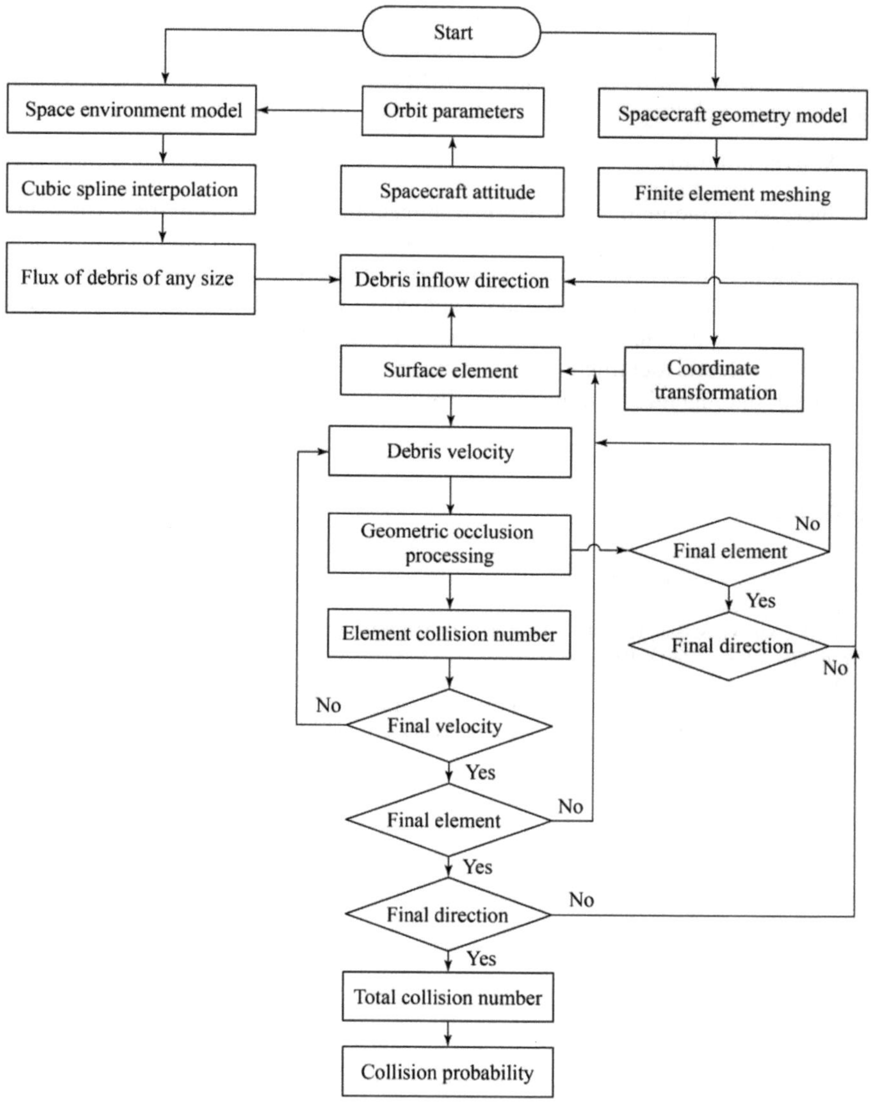


**Fig. 4.8** Program of space debris collision module

continuous improvement of debris data accuracy has laid a solid foundation for the establishment of a more accurate debris environment model that makes accurate early warning possible.

2. **Comprehensive Warning**

There are many methods for debris collision warning. A spacecraft can be warned against a fragment by calculating the probability of collision between them or by determining the distance between them. The different warning methods have their own advantages and disadvantages while complementing each other. To provide more accurate collision warning, multiple warning methods can be combined to improve the warning reliability and minimize unnecessary maneuvers in orbital transfer.

3. **Active Avoidance**

The increasing amount of space debris is threatening our space activities. The debris warning is built upon a variety of calculations using a huge amount of debris data, but the warning results often have a large error. Therefore, the passive protection of spacecrafts against small debris can be combined with automatic identification and avoidance of the threatening debris. We believe that in the near future, the technology of automatic debris identification and avoidance will be used in space activities with the further intelligentization of spacecrafts.

## References

1. Zhang, R., & Yang, K. (2017). *Spacecraft flight collision warning technology*. National Defense Industry Press.
2. He, A., Xu, K., & Bao, K. (2014). Design of debris collision warning system. *Command, Control and Simulation, 36*(5), 105–108.
3. Wang, H., Yu, Q., & Liu, Y. (2008). Risk assessment system for space debris collision. *Journal of Beijing Institute of Technology, 28*(12), 1039–1042.
4. Wang, L. (2014). Research status and development trend of space debris warning. *Consumer Guide, 2*, 36.

# Chapter 5
# Structural and Material Protection Against Space Debris

## 5.1 Introduction

The average relative velocity between a micrometeoroid and a spacecraft can reach 19–22 km/s, and the average relative velocity between space debris and a spacecraft can reach 10 km/s. The micrometeoroids and space debris seriously threaten the safety of the orbiting spacecrafts. The space debris protection can be traced back to the 1950s, when the exploration of effective protective structures based on ultra-high speed collision was started and extended to the "Apollo" program in the 1960s in order to ensure the safety of intercontinental ballistic missiles and deep space probes and to understand the damage effect of debris and meteorite collision on spacecrafts. In recent decades, the impact of space debris has become more and more serious. On February 11, 2009, "Iridium 33", an American commercial communications satellite, collided with the obsolete Russian military reconnaissance satellite "Cosmos 2251" 789 km above Siberia at a relative speed of 11.6 km/s, generating a large amount of space debris. The event attracted intense attention from the international community. On July 24, 1996, the French military reconnaissance satellite "Cerise" collided with the remains of the carrier rocket "Ariane" 66 km above the ground at a relative speed of 14 km/s and got out of control. The International Space Station (ISS) is the largest spacecraft that has been built and is also most seriously affected by space debris. By the end of 2018, the ISS had made more than 26 orbital maneuvers to avoid the impact of catalogable space debris[1] The ISS is hit by space debris almost every year. The British Space Agency detected the 3–7 mm craters/holes on the early Hubble Telescope's solar wing, which were caused by 5000–6000 M/OD impacts including 150 impacts that completely penetrated the solar wing. A spacewalk in 1999 photographed 571 small impact craters on the surface of the Hubble Telescope. The Hubble Telescope has 34,000 debris-impact craters, including 5000 craters larger than 0.5 mm.

[1] Li Ming. Frontiers and Prospects of Space Debris Environment Governance [C]. The 10th National Space Debris Academic Exchange Conference, 2019.



J. Zhang et al., *Space Debris Research*, https://doi.org/10.1007/978-981-96-9570-6_5

Space debris is potentially harmful. According to its size, space debris can be divided into large debris, small debris and micro-debris. Among them, the debris with a size above centimeter level is large debris or catalogable debris; that below centimeter level is called micro-debris; and that between large debris and micro-debris is called small debris. By the end of 2019, the total mass of the orbiting objects was 8400 t. The tracked and cataloged Earth-orbiting fragments included 23,000 fragments larger than 10 cm, about 750,000 fragments within 1–10 cm, about 100 million fragments within 1–10 mm, and tens of billions of fragments smaller than 1 mm [1]. Large debris can be detected and tracked through the monitoring network. When a collision is detected, the spacecraft will make an active maneuver to avoid it. The amount of small debris is relatively small. The detection and prevention of small debris is still a major challenge in the space field. Because the small debris has small kinetic energy during a collision, the spacecrafts generally take passive protective measures against it. With the improvement of debris-monitoring and protection capabilities, the management and prevention of small debris will be gradually enhanced. The micro-debris has a huge amount and cannot be tracked, positioned and cataloged one by one. It can lead to cabin wall perforation, radiator tube perforation, battery damage, battery performance decline, and photosensitive devices degradation. Like small debris, the micro-debris can only be prevented by passive protective measures.

The harm of space debris impact to the spacecraft subsystems, mainly including structural parts/components and functional parts/components [6], is described below. For the structural systems (such as solar panel), the debris impact will cause the cratering, perforation and cracking of solar panel surface and the spalling and peeling of the panel back. The electromagnetic pulse generated by the collision will lead to abnormalities in the electronic system. In addition, the solar array, battery and distribution network may be damaged by the collision, resulting in the failure of the energy system. The collision will also cause the performance degradation or failure of optical system (such as the performance of photosensitive devices). The storage tank and pipeline of the spacecraft will leak and burst due to the collision, causing the failure of propulsion system. The surface coating of thermal control system will also fall off, resulting in the decline or failure of thermal sensitive function and thermal protection function. In Table 5.1, the damage effects of the debris of different sizes on the equipment/structures of satellites are described and also indirectly reflect the requirements for satellite protection.

To ensure high reliability, long life and safe operation of the spacecraft in the space environment, appropriate protective measures (for example, flight attitude adjustment, equipment layout adjustment, use of the materials with strong protective performance, and protective structures adding) shall be taken to improve the survival probability of the spacecraft in the micro-debris environment. Meanwhile, the increase of non-payload mass arising therefrom shall be minimized. The design of spacecraft protection shall take fully account of the cost, manufacturing process, design scheme implementability and many other factors, follow the design principles

**Table 5.1** Debris impact to satellites

| Equipment/structure | Influencing particle size | Damage form | Damage effect |
|---|---|---|---|
| Honeycomb panel/ aluminum panel | Millimeter level | Cratering and perforation | The structural load will not be affected, but the perforation will affect the internal equipment |
| Multilayer insulation material | Below millimeter | Perforation or cumulative damage | The thermal control effect will be affected |
| Star sensor, earth sensor and their hood | Millimeter level | Cratering and perforation | The Perforation will cause stray light invasion, and the secondary debris clouds will contaminate the lens and affect its normal operation |
| Camera and its hood | Millimeter level | Cratering and perforation | The Perforation will cause stray light invasion, and the secondary debris clouds will contaminate the lens and affect its imaging quality |
| Radar and other antennas | Millimeter level | Cratering and perforation | The normal operation will basically not be affected |
| Data bus and power bus | Below millimeter | Bus disconnection | Equipment failure and even satellite failure |
| Fuel pipeline | Millimeter level | Cratering and perforation | Pipe blockage and fuel leakage |
| Battery | Millimeter level | Cratering and perforation | Electrolyte leakage and battery explosion |
| Storage tank | Millimeter level | Cratering and perforation | Propellant leakage and tank explosion |
| Inboard electronics | Millimeter level | Cratering and perforation | Equipment failure and even satellite failure |
| Solar wing | Millimeter and micrometer levels | Cratering, perforation or cumulative damage | The perforation will result in the loss of battery power-generation capacity, and the cumulative damage will affect the power generation efficiency of batteries |

of "making full use of external layout" and "minimizing the mass of special protection", and consider the trade-off between outer envelope and protection structure mass.

## 5.2 Space Environment Effects

The design of spacecraft protection structure shall consider not only the inherent properties of the material, but also the effects of space environment factors on the material properties. In addition to space debris, the space environment factors such as vacuum, high and low temperature, space radiation, micrometeoroids, plasma, planetary atmosphere and magnetic field[2] will produce various complex effects on the surface materials and internal devices of the spacecraft and even cause their irreversible damage, for example, radiation damage, mass damage, vacuum discharge and molecular pollution. Some common space environments and their effects are described below.

1. **Vacuum**

In the Milky Way galaxy, the interplanetary atmospheric pressure is $10^{-13}$ to $10^{-18}$ Pa, which indicates an ultra-high vacuum state. Therefore, the vacuum effects should be considered in this environment. For example, the pressure difference effect will cause the internal deformation or damage of the sealed cabin or protective structure, which may lead to the leakage of protective structure interlayer and the decline of protection performance in vacuum environment. The vacuum discharge effect will cause the metal erosion due to the emission of secondary electrons, which, in turn, will lead to the temperature rise and the rise of nearby gas pressure so that the protective structure will be affected to some extent. Under the vacuum gas effect, the gas originally absorbed on the material surface or dissolved inside the material or some permeable gas will diffuse out in vacuum environment, causing molecular pollution and changing the properties of the material coating. The evaporation, sublimation and decomposition effects of the material will cause the material component changes and material mass loss and change the original properties of the material such as thermal physical properties and dielectric properties. In addition, the uneven sublimation of spacecraft surface material will lead to certain surface roughness, the possible changes in the interface between the material inside and outside in high vacuum and then the changes in mechanical properties of the material. The lack of oxidation film or other surface protection films during evaporation may lead to the changes in the adaptation coefficient and emissivity of material surface and the significant changes in mechanical properties, creep strength and fatigue stress of the material. In terms of thermal radiation effect, the spacecraft heat is transferred to the outside mainly through radiation in vacuum environment, so the radiation characteristics of the

[2] Han Zengyao, Cai Zhenbo, Liu Zhengshan. Some Problems about Space Environment Effects on Spacecrafts [C]. The 14th National Conference on Physics and Mechanics.

spacecraft surface play an important role in the control of spacecraft temperature. The design of protective structure must consider the effect of the heat transfer dominated by radiation and conduction in vacuum environment in order to prevent the heat entrapment due to the blocking of external protective structure.

2. **Alternating High/Low Temperatures**

In space environment, the mechanical properties of onboard protective material are seriously affected by alternating high and low temperatures. When the spacecraft is in orbit, its protective structure is exposed to repeated high-low temperature cycles due to the change of spacecraft attitude, which result in the changes in the mechanical properties of spacecraft surface material. In addition, the frequent thermal cycling will produce micro-vibration to affect the strength of the protective structure. With the increase of the spacecraft orbiting time, the protective material will be exposed to more temperature cycles and will be more seriously affected by them. Therefore, the design of protective structure must consider the effect of high-low temperature cycles on the physical properties, strength and life of the material and structure.

3. **Space Radiation Environment**

The space radiation environment includes trapped radiation belts, solar cosmic rays, galactic cosmic rays and others. The trapped radiation belts refer to the areas of charged particles around planets trapped by the planets' magnetic fields, such as the Earth's radiation belt, Jupiter's radiation belt and Saturn's radiation belt. They are mainly composed of protons, electrons and a small number of heavy ions. The solar cosmic rays mainly refer to the high-energy protons, electrons and heavy nucleon flows emitted during solar flares, among which the protons have a largest percentage and a general energy range from 10 MeV to tens of GeV. The galactic cosmic rays (Fig. 5.1) are the charged particles coming from the outside of solar system, which have high energy and low flux and consist of 85% protons, 14% alpha particles, and 1% heavy ions. The energy of galactic cosmic particles ranges from tens of MeV to $10^{14}$ MeV.

The exposure of a manned spacecraft to space radiation will cause damage to the genetic material in the astronauts' tissues or organs, in the forms of DNA double-strand breaking, gene mutation and chromosomal aberration, which may lead to cell inactivation and genomic instability so as to destroy the human bone marrow, skin, central nervous system, reproductive system and other tissues, organs or systems. In addition, the damage of space radiation to the human body will accumulate over time. The galactic cosmic rays have high radiation energy and a penetration ability too strong for an ordinary mass thickness to shield, thereby resulting in more serious radiation damage. The above space radiation particles will not only affect the surface material of the spacecraft, but also form a radiation environment inside the spacecraft to change the performance of some materials and devices and even cause their permanent damage. The electrons, protons, heavy ions, photons and other space radiation particles can produce the single event effect, total dosage effect, surface charging/discharging effect, displacement damage effect and internal charging effect

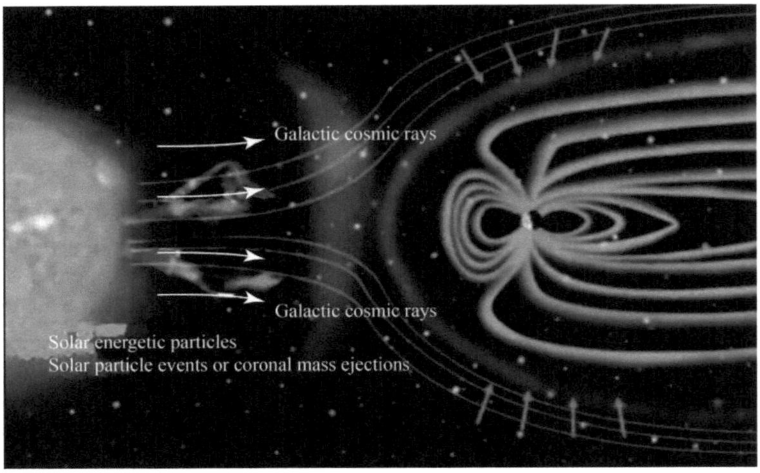


**Fig. 5.1** Schematic of galactic cosmic radiation (see the colored illustration)

on spacecraft materials or components. In particular, solar batteries, optical surfaces, organic materials, semiconductor devices and integrated circuits are vulnerable to space radiation damage mainly by ionization and displacement[2]. Ionization is the process in which the incident particles induce the ionization of target atoms and the excitation of extranuclear electrons in the material to form the electron hole pairs. The ionization can cause the degradation of device performance and produce single particle effect and total dosage effect. The interaction between photons (especially high-energy photons) and polymer materials will cause the chemical bond breaking or cross-linking of polymer materials, thus resulting in the degradation of physical or chemical properties of the materials. Displacement is the departure of target atoms from their original positions and the formation of interstitial atoms and vacancies due to the interaction and kinetic energy exchange between the incident particles and the material atoms. In addition, the displaced atoms may collide with other atoms, forming a displacement chain. The interstitial atoms or vacancies generated by displacement generally have strong activity and are the sources or traps of carriers in semiconductor materials or devices.

4. **Micrometeoroids**

A micrometeoroid (Fig. 5.2) is similar to, but different from, space debris. The difference lies in the fact that the former is an object that naturally exists in space, while the latter is a man-made result. They both have similar effects on spacecrafts and have high speeds. Due to high kinetic energy, their collision with spacecrafts may cause the loss of some of the spacecraft functions or even the complete destruction of spacecrafts, posing a great threat to manned spacecrafts and even threatening the lives of astronauts. Although the probability of a collision between a micrometeoroid and a spacecraft is much lower than that between space debris and the spacecraft due

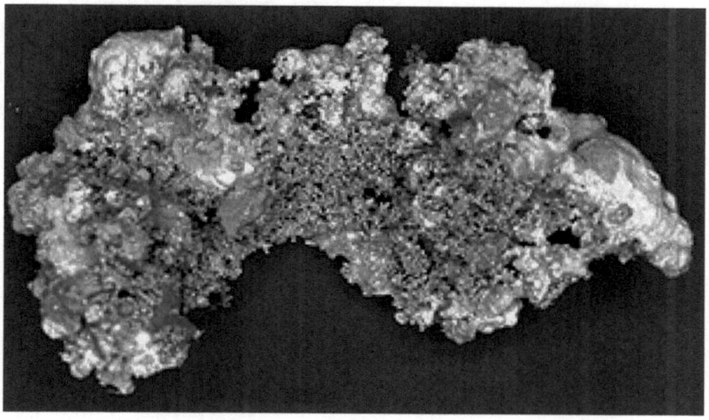

**Fig. 5.2** A micrometeoroid falling to the Earth

to the influence of the Earth, the risk posed by micrometeoroid environment to the spacecrafts still needs to be assessed [3].

5. **Plasma**

Plasma is the basic material of the universe, filling the whole space from the ionosphere just outside the Earth's atmosphere, to the Earth's magnetosphere, the sun, the heliosphere, and even interstellar space, nebula and star clusters. Space plasma is often found in the auroral belt (Fig. 5.3) and the polar cap region especially during solar maximums. The plasma environment will affect the spacecraft operation, and will produce the charging effect, radiation effect, dynamic effect and pollution effect, especially under the condition of geomagnetic subexplosion. Depending on the energy of charged particles, the effects of space plasma on spacecraft surface mainly include radiation damage, surface charging and current coupling [4].

Among the space environment factors mentioned above, the space debris, micrometeoroids, high and low temperature, space radiation environment, plasma and vacuum have a particularly significant effect on the performance of spacecraft protection structure. However, due to the limitation of simulated environmental conditions (such as high/low temperature and radiation), the impact of space environment factors is not considered in the data of high-speed impact damage and protection characteristics of most of the protective structures. Therefore, in the development process of space protection technology, the analysis of protective material performance against space environment factors will be gradually improved to obtain a more reliable spacecraft protection design.

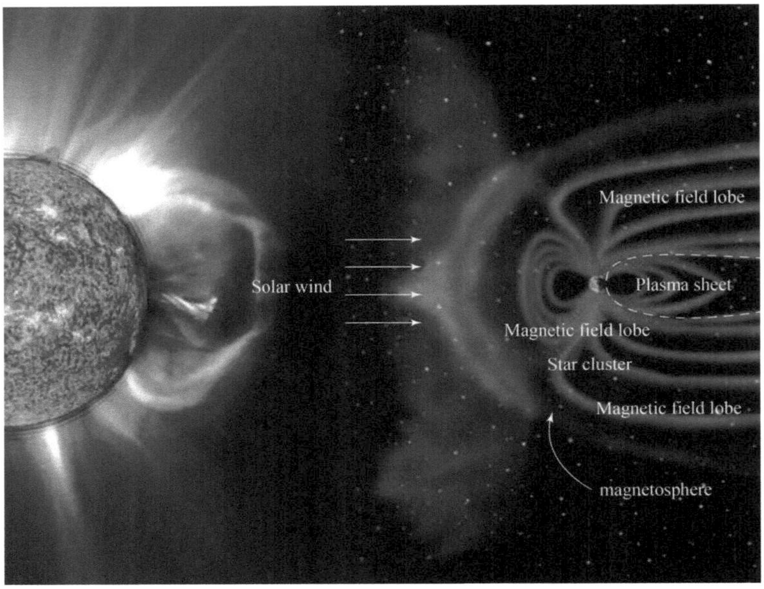


**Fig. 5.3** Plasma in the northern lights (see the colored illustration)

## 5.3 Theory of Protection

In a broad sense, the space debris protection includes the measures for improving the spacecraft's survivability in the space debris environment, such as: (1) improving the spacecraft vulnerability to debris impact; (2) assessing the spacecraft collision risk based on a debris environment model; (3) optimizing the onboard equipment layout based on collision risk assessment; and (4) adding the protective shields to the spacecraft areas with high collision risk and vulnerability. The protection against space debris can be divided into passive protection and active protection.

The passive protection is mainly a type of structural protection against the ultra-high speed collision with millimeter- and micron-sized M/OD smaller than 1 cm. The spacecrafts with different missions and the different subsystems of the same spacecraft have different passive protection requirements due to their different damage/failure modes. For manned spacecrafts (such as space stations), a high safety probability is required (0.998 for ISS), and the protection is focused on the crew safety and the avoidance of system-level failure. Therefore, the design of protective structure for a manned spacecraft is mainly to provide effective and ingenious shielding protection for key cabins (e.g. pressurized cabin) and most of the sensitive devices according to the protection grade requirements. However, for non-manned spacecrafts (e.g. satellites), the use of a special protective wall is usually avoided due to the complexity of their failure mode and the strict requirements for their quality. Instead, their exposed structures and key components are designed with protective measures (other than protective shields), which can be roughly divided into main structure protection, composite material protection and structural occlusion protection.

The basic theory of passive protection is that the initially incident debris collides with the protective shield installed on the exterior of cabin wall and is then fragmented, melt and even vaporized into a secondary debris cloud, so as to minimize and disperse its kinetic energy, significantly reduce the impact energy flux density on cabin wall and weaken the damage to the cabin wall. But if the spacing between the shield and the cabin wall is not large enough, even the resultant secondary debris cloud may still cause the perforation, spalling, peeling and cratering of the cabin wall because of the failure of effective debris cloud diffusion before collision.

The active protection against space debris is to use a detector to monitor the debris approaching the spacecraft, predict its impact position and time, and quickly release and deploy the protective barrier at the impact position with the help of the control system, so as to "shut out" the debris under the threat of collision. Like the passive protection, the active protection also relies on a multi-layer protective shield to decelerate the collision speed of the incident debris and reduce its damage to the cabin wall. The only difference is that the active protection shield is composed of multiple flexible layers filled with foam or gas to meet the requirement of collapsible deployment and achieve the purpose of protecting the spacecraft. The active protection mechanism can be released directionally, recycled, and folded after daily use. Compared with passive protection mechanism, the active protection mechanism has the advantages of smaller size, smaller mass and higher benefit–cost ratio, but still has many engineering problems to be solved because active protection is a new concept. So far, the active protection has not been widely used, thus having great significance in the future research and application.

Thus, although different protective structures have different forms, they all follow the common design idea of "protective shield + cabin wall + their spacing". The material properties and parameters of a protective structure determine its protective performance. In addition to meeting the requirement of small mass, a protective structure with excellent performance should ensure the full melting or vaporization of the incident debris after its collision with the shield, the generation of as few and small re-sputter particles as possible, and the full diffusion of the secondary debris cloud before colliding with the spacecraft body. The spacecraft body material is also required to have a strong ability to absorb micro-debris particles.

## 5.4 Advanced Protective Materials

The long-time exposure to harsh space environment will produce many adverse effects on the material properties of a spacecraft. For example, atomic oxygen corrosion will lead to the degradation of material properties; space radiation will degrade the performance of organic materials; thermal cycling will reduce dimensional stability and mechanical properties of the materials; the ultra-high speed collision of space debris will cause mechanical damage and destruction to the materials;

ultra-high vacuum environment will lead to the decomposition, disintegration and gassing of organic materials. Therefore, from the perspectives of space environment, flight mission, launch capacity and cost–benefit ratio, the spacecraft protection design should choose the aerospace materials with high strength, good quality, high radiation resistance, stable chemical properties and high protection efficiency. The characteristics of these materials will have a great impact on their protection ability. According to the space environment types and protection requirements, the characteristics of protective materials are summarized in Table 5.2.

According to the existing protection designs, the material of buffer shield should generally have a high impact impedance to fully break the low-velocity particles, and the material of rear wall should have a high transverse sound velocity to fully diffuse the impact pulses of a debris cloud. The selection of protective structural materials mainly includes the selection of rear wall material and more importantly, the selection of buffer shield material. The general principle for selecting the buffer shield material is that the chosen material should have a small mass, a strong ability to break the incident particles and a small threat of its fragmentation to the rear wall and should ensure a large diffusion angle, low expansion speed and small secondary re-sputter effect of the debris cloud.

The existing spacecraft protection materials commonly used at home and abroad can be divided into metal materials, metal composite materials, fiber materials, insulation materials and ultra-high temperature materials. Among them, aluminum alloy is the metal material most commonly used in the design of protective structures,

**Table 5.2** Summary of the characteristics of protective materials

| Requirements for characteristics | Contents |
|---|---|
| Light weight | To improve the payload mass of spacecraft, reduce the proportion of structural mass in the spacecraft, and choose light materials |
| Mechanical properties | The structural materials are required to have high strength, high modulus and good ductility, and had better have high specific modulus and specific strength |
| Physical properties | The structural materials should have a small coefficient of linear expansion, high specific heat and high thermal conductivity to ensure uniform temperature distribution in the structure and avoid too high temperature stress and deformation |
| Space environment resistance | The structural materials of a spacecraft orbiting for a long time, especially the outboard structural materials directly exposed to space, are required to have good stability in space environment (such as irradiation resistance and aging resistance) |
| Material gassing in vacuum | In vacuum environment, the gassing of a material will not only degrade the performance of the material, but more importantly, contaminate nearby optical, thermal or electrical devices. Therefore, the requirements of vacuum gassing limiting should be put forward for the spacecraft's structural materials. Generally, the total mass loss of the materials is not more than 1%, and the volatile condensate is not more than 0.1% |

mainly because of its low density, high strength, good corrosion resistance, easy machining, and an affordable cost that can meet the needs of the engineering budget to a large extent. Aluminum alloy is heavy and thus is mostly used as the material of rear wall and buffer layers in the spacecraft protection design solutions. An effective protective aluminum mesh is usually installed in front of the outermost buffer wall of the spacecraft protection structure to break the incident space debris. Metal composite material is a new type of composite material made by blending the fiber particles into a metal, and has high shear strength, excellent resistance to fatigue and heat, and a very small coefficient of thermal expansion. Aluminum honeycomb core material is widely used in spacecraft protection design. In an ultra-high speed oblique collision, this material is very helpful in distributing the impact energy over a large area of the backboard to reduce the probability that the spacecraft body will be penetrated. When facing the tunnel penetration effect in a head-on collision, aluminum honeycomb sandwich can effectively limit the expansion of secondary debris cloud and confine the impact effect to a small area. In addition, with the development of material technology, advanced ceramic fiber, particle reinforced aluminum matrix composites, foamed aluminium [5] and other protective materials will have a very good application prospect in aerospace. Fibrous materials are widely used because of their high modulus, high strength and low mass. High-strength Nextel fabric is a kind of ceramic fiber woven fabric. In a high-speed collision, the aluminum/boron/silicon fibers in Nextel fabric can propagate strong shock waves into the incident debris to break it into small fragments, while producing a large number of small fibers with basically no threat to the bearing layer so as to significantly reduce the collision threat to the spacecraft. Kevlar is an aromatic polyamide woven fabric, as shown in Fig. 5.4. The material has low density, high strength, good toughness and high-temperature resistance, and is easy to machine and form. Its strength is 5 times that of steel of the same mass, and its density is only 1/5 of steel. Kevlar is widely used in the field of armor protection. Kevlar can very effectively transfer the collision momentum of small particles, and quickly redistribute the momentum of particles along the direction of fiber length. The combination of Nex-tel and Kevlar can fully reduce the damage effect of various incident fragments in a collision. Beta fabric is a tightly woven fiberglass fabric often covered with a layer of Teflon coating. The fabric is mainly used in multilayer insulating materials to provide thermal protection for spacecraft components, resist the material degradation caused by atomic oxygen invasion, and slow down the speed of the incident space debris to some extent.

Basalt fiber is a continuous fiber that is made by using the platinum–rhodium alloy wire-drawing bushing to quickly draw the molten basalt material at about 1500 °C, as shown in Fig. 5.5. Its performance is between high-strength S glass fiber and alkali-free E glass fiber. Pure natural basalt fiber is generally brown and somewhat golden. Basalt fiber is a new type of inorganic green high-performance fibrous material, and is formed by using a special device to quickly draw the molten basalt material that is made of silicon dioxide, alumina, calcium oxide, iron oxide, titanium oxide and other oxides. The basalt fiber has not only good stability, but also electrical insulation, corrosion resistance, combustion resistance, high-temperature resistance and other excellent properties. This continuous fiber has been widely used

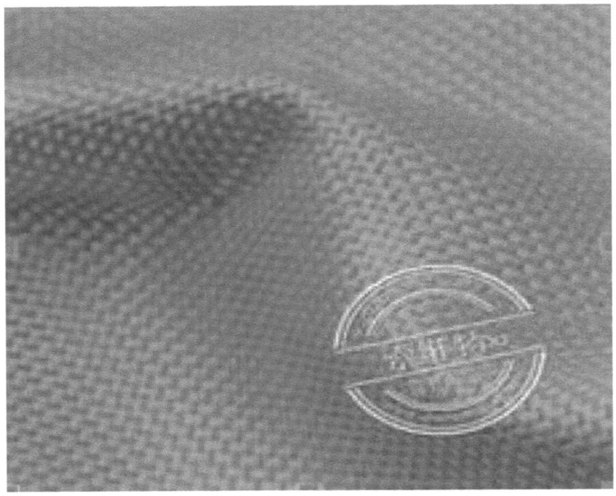

Fig. 5.4 Kevlar fiber fabric (see the colored illustration)

in fiber reinforced composite materials, friction materials, shipbuilding materials, heat insulation materials, automotive industry, high-temperature filter fabrics and protection field.

To adapt to the abnormal temperature environment in space, thermal insulation materials are widely used in space engineering. For example, the low-density ablative materials based on phenolic aldehyde, epoxy resin or silicone rubber are suitable for heat protection and insulation at higher temperatures, and can ensure the performance and temperature of protective structures. Ultra-high temperature materials are able to keep their physical and chemical properties stable when the spacecraft passes

Fig. 5.5 Basalt fiber fabric (see the colored illustration)

through the atmosphere in the high-temperature space environment. These materials can play a good role in spacecraft protection.

In addition to the above materials used for heat insulation and collision protection, the selection and structural design of radiation-hardened materials to improve the radiation resistance of spacecrafts are also indispensable due to the serious threat of space radiation environment to the safety and reliability of the orbiting spacecrafts. Especially in the process of material selection, the mass shielding needs to consider the integrated influence of primary radiation and secondary radiation, as well as the limitation of payload increase to the increase of structural mass. Therefore, the development of the composite materials composed of low atomic-number elements and high atomic-number elements that can not only reduce the structural mass but also improve the radiation resistance is one of the important development directions of spacecraft protection materials. In addition, the application of nanotechnology in radiation-hardened materials is also a hot spot of spacecraft radiation protection research. Figure 5.6 shows a new anti-radiation nanomaterial developed by the Australian National University. The radiation-shielding layer of the nanomaterial can provide the same protection as a thicker radiation-shielding material. The nano-coating can reflect the light and rays of specified wavelengths, including far-infrared and ultraviolet rays. At different temperatures, the new nanomaterial can reflect different wavelengths by changing the refractive index of its nano-coating through the electric heating or cooling of the material. The overlay of multiple layers of this material can shield all kinds of light and harmful rays. This material can be used to make new ultra-thin space suits or serve as the protective structure for spacecrafts.

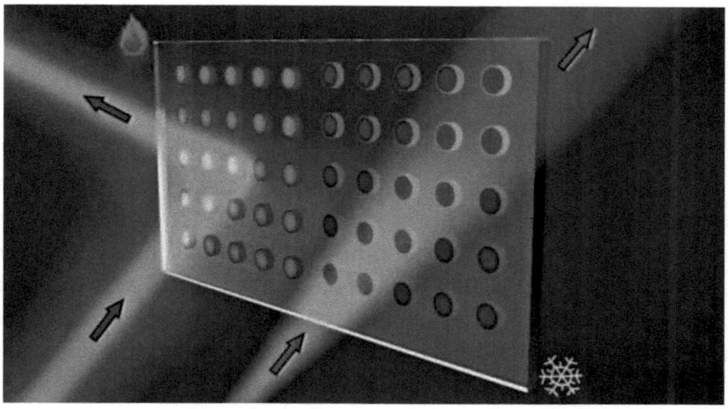

**Fig. 5.6** New anti-radiation nanomaterial (see the colored illustration)

## 5.5 Shield Configuration

In order to ensure the safety of spacecrafts (especially the large spacecrafts flying in low Earth orbit for a long time), the protective structure against micro-debris must be designed to improve the survivability of the spacecrafts. The fundamental task of passive protection design is to combine and arrange the selected protection materials in a specific way to achieve the required protection performance. From the perspective of configuration, each type of protective structure is composed of three parts: the rear wall, one or more buffer shields, and the distance between the rear wall and the outermost buffer shield. Therefore, the parameters of the designed protective structure include the distance between the outermost shield and the cabin wall, the distance between shields, the thickness of each shield, and the orientation and position of the supporting ribs. Shield is an important factor in spacecraft protection design, and is closely related to spacecraft mass and launch cost. In addition, the design of spacecraft protection structure and the design of payload are mutually constrained.

Depending on the number of protective shields, the existing protective structures can be classified into three types: single-shield structure, double-shield structure, and multiple-shield structure. Among them, the single-shield structure has only one shield, and is simple in structure and low in protection performance. The single-shield structures are mainly divided into two types: basic Whipple structure and foamed-aluminum shield structure. So far, four types of shields have been listed in the IADC's "Protection Manual", including: Whipple shield, stuffed Whipple shield, multi-shock shield, and mesh double bumper shield.

### *5.5.1 Whipple Shield*

In the 1940s, Fred Whipple first proposed a protective configuration with a buffer shield, later called Whipple shield (Fig. 5.7), which was composed of an aluminum shield (bumper shield), a rear wall (protected structure), and the spacing between them. The shield is used to break the coming M/OD. If the collision velocity is high enough, the broken particles will be melted or even vaporized to form a debris cloud after the collision with the Whipple shield, so that the kinetic energy of the incident debris will be partially absorbed and highly dispersed to reduce the damage to the rear wall. In about the 1990s, a variety of shields were developed internationally based on Whipple shield to improve the protection performance. Among them, the relatively mature configurations include stuffed Whipple shield, mesh double bumper shield and multi-shock shield.

When the incident debris collides with the Whipple shield, the material will be fragmented, melted and even vaporized into plasma, forming a secondary debris cloud. Then, the secondary debris cloud moves and diffuses towards the cabin wall, resulting in a gradual decrease in its energy flux density, which significantly weakens

Fig. 5.7 Whipple shield (see the colored illustration)

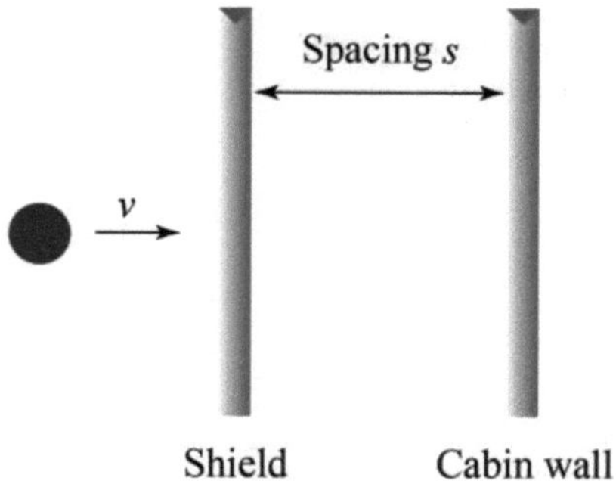


the later impact damage to the cabin wall. A large number of ground-based hypervelocity collision simulation tests showed that the protecting effect of Whipple shield improved with the increase of incident projectile velocity. When the incident velocity was lower than 2 km/s or in a oblique direction, the protection performance declined significantly and the protecting effect was not ideal. The Whipple shield was widely used in the early protection designs against space debris, such as American "Apollo" spacecrafts, "Skylab" space station, and Russian "Salyut" space station.

### 5.5.2 Stuffed Whipple Shield

The main difference of stuffed Whipple (SW) shield from a simple shield is the addition of an intermediate shield between the outer shield and the rear wall. The additional shield is usually a sandwich panel composed of several layers of Nextel and Kevlar fabrics, and can capture, slow down and further fragment, melt and vaporize the secondary debris cloud generated by the collision between space debris and the outer shield, so as to further improve the efficiency of cabin wall protection. The structure of SW shield is shown in Fig. 5.8.

### 5.5.3 Mesh Double Bumper Shield

The mesh double bumper (MDB) shield consists of three bumper shields and a rear wall, as shown in Fig. 5.9. The aluminum mesh is made of orthogonally woven aluminum wires. The continuous bumper shield is made of aluminum alloy. The composite bumper shield is made of Spectra, Kevlar or Nextel fabrics. The cabin wall is generally made of aluminum alloy. Compared with the Whipple shield with the same protection performance, the MDB shield has a smaller mass (about 50% of the Whipple shield) and produces a smaller amount of destructive secondary debris. The ROS section of the International Space Station is just protected by MDB.

Fig. 5.8 Structure of stuffed Whipple shield (see the colored illustration)

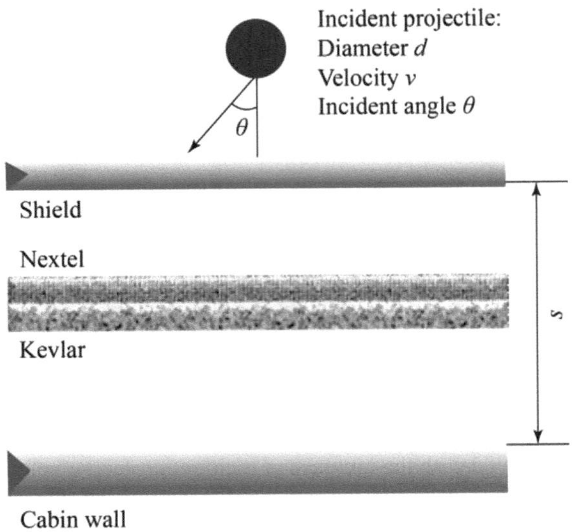


Fig. 5.9 MDB structure (see the colored illustration)

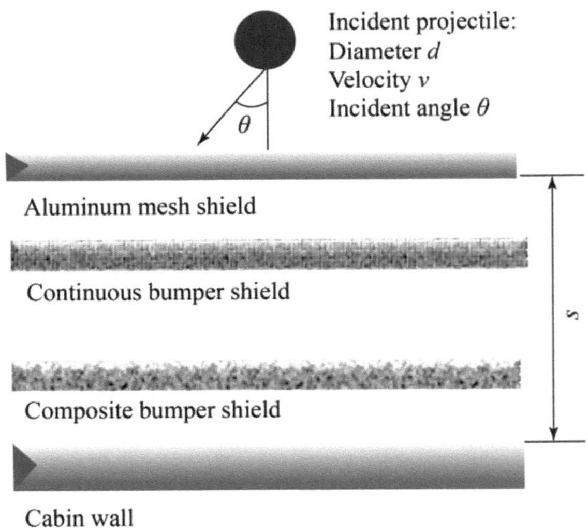


### 5.5.4 *Multi-shock Shield*

The main difference of multi-shock (MS) shield from a simple Whipple shield is the use of several (generally 3–5) Nextel shields with the same spacing instead of only one shield. The repeated impact effect of the Nextel shields allows the incident debris and the secondary debris cloud to be continuously broken, melted and vaporized, thus significantly reducing the damage caused by the incident debris to the cabin wall.

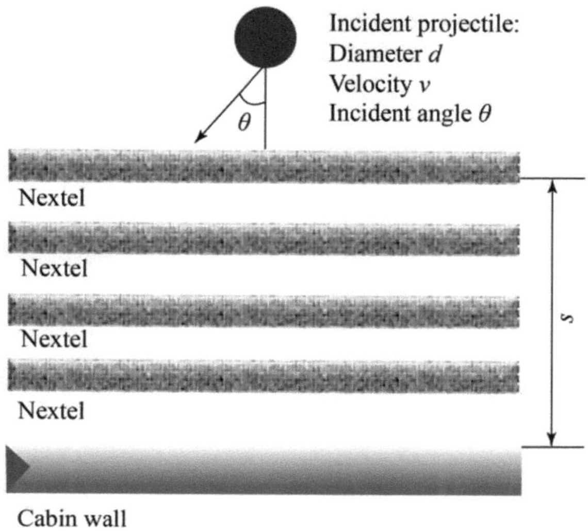


**Fig. 5.10** Structure of MS shield (see the colored illustration)

The MS shield can not only significantly improve the protection efficiency and effectively reduce the mass of the protective structure, but also can be used with other shield types. Therefore, it is now commonly used in spacecraft protection design. The structure of MS shield is shown in Fig. 5.10.

### *5.5.5 Other Shields*

In addition to the above four types of shields included in the IADC's "Protection Manual", many shield designs have been put forward or are under test, such as Hybrid Nextel/Aluminum Multi-Shock Shield (HMS), Corrugation Stuffed Whipple Shield (CSW) and Aluminum Double Bumper Shield (ADB). The feature of HMS is that two or three Nextel shields are added to the Whipple shield to provide the same protection efficiency as the MS and Whipple shields, as shown in Fig. 5.11.

The CSW shield (Fig. 5.12) is formed by corrugating the outer shield of the SW structure, and is aimed to improve the protection efficiency by increasing the incident angle of space debris. The feature of ADB shield (Fig. 5.13) is the addition of an aluminum shield between cabin wall and Whipple shield, with the aim of improving the protection efficiency through the deceleration effect and repeated impact on the incoming debris.

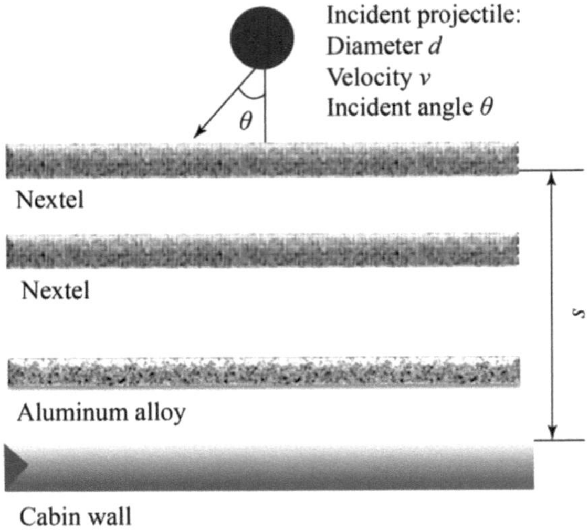


**Fig. 5.11** Structure of HMS shield (see the colored illustration)

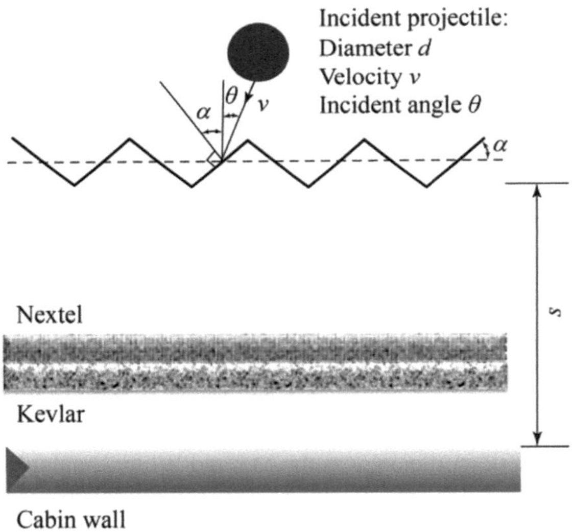


**Fig. 5.12** Structure of CSW shield (see the colored illustration)

## 5.5.6 *New Shields*

As can be seen from the above description, the existing protection designs are basically the passive shields based on the Whipple shield structure. Their application is mainly focused on large spacecrafts (such as manned spacecrafts and space stations), and is limited by mass, volume, launch cost and other factors to only the effective protection against the impact of millimeter-sized debris. So far, almost all of Chinese

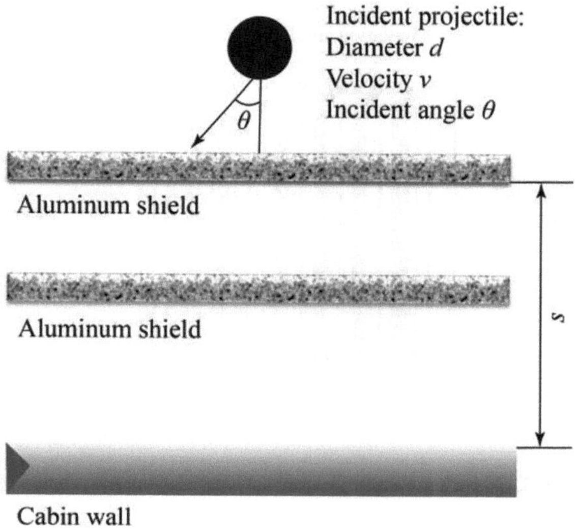


**Fig. 5.13** Structure of ADB shield (see the colored illustration)

satellites have not been protected by shields. However, with the deterioration of space debris environment and the development of American kinetic-energy collision weapons, the spatial survival environment is becoming more and more harsh. Under this background, an important development direction of space debris shield is to develop an active protective payload to replace or complement the traditional externally-mounted passive protection structure so as to realize the combination of active and passive protections.

Due to the limitation of size and mass, the above common protective structures can only withstand the impact of a piece of space debris smaller than 1 cm. Although the flexible multi-layer HMS structure can greatly improve the protective performance, its application is still limited by the optimal shield spacing more than 30 times of the projectile diameter, which is impossible due to the limitation of the onboard payload space. In recent years, the concept of inflatable and deployable capsule has brought us enlightenment in debris shield design. A multi-layer flexible protection system that can be rapidly inflated and deployed has been developed (Fig. 5.14) and can be launched into space in the form of satellite payload. When a piece of debris comes, the protection system is quickly released by the release mechanism and the rapid inflation system is triggered. Then the multi-layer flexible housing is deployed rapidly under the action of high pressure gas. By adjusting the direction of release, a large-spacing multi-layer flexible protective structure is formed at the place where the target with kinetic energy is coming, so as to realize active effective protection against the debris. After the completion of one protection task, the system can be reused. The simulation test based on the existing material technology shows that [6] the inflated and deployed protection structure can effectively protect the cabin wall in space from the impact of a high-speed projectile with a diameter of 1–2 cm, and can meet the functional design requirements.

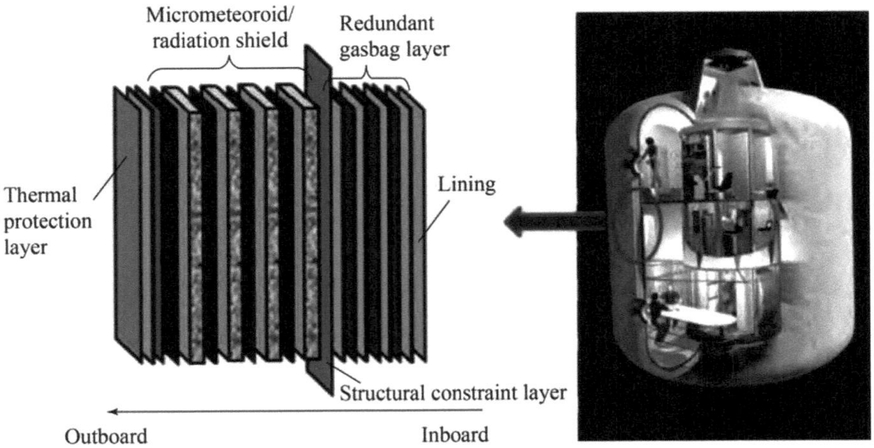


**Fig. 5.14** Structure of a flexible capsule housing (see the colored illustration)

The system should meet the following functional design requirements:

(1) Detection and identification of centimeter-sized targets. The on-orbit identification of centimeter-sized space debris and the determination of its parameters provide an important basis for spacecraft orbit warning and active debris avoidance. Accurately judging the collision risk and giving an instruction to the release mechanism is the key to giving play to the role of active protection system.
(2) Design of debris shield function. The protective structure should be able to break the incident debris into small particles, and allow the debris cloud generated by the collision to fully diffuse. When the debris cloud collides with the cabin wall, the active protection system can reduce the impact energy of the debris cloud per unit area, so as to supplement the protective capacity of passive protection structure and prevent the damage of space debris to the high-risk parts of the space station.
(3) Design of folding and deployment functions. To meet the requirement of large station area protection and ensure a reasonable and feasible launch envelope, the protective structure should provide the functions of in-launch folding and on-orbit deployment. To increase the folding rate of the protective structure and reduce the complexity of the deployment mechanism, the protective structure should be made of a flexible composite material that can be folded flexibly for many times and can be inflated and deployed in orbit.
(4) Functions of fast inflation and release. A structure for rapid inflation and deployment should be designed to ensure that the best opportunity for shielding the spacecraft from debris impact will not be missed due to slow release speed.
(5) Functions of retraction and redeployment. The protective structure has a large area when being deployed. To reduce the resistance to spacecraft flight and the energy consumption in orbit maintenance and avoid the occlusion of other

onboard devices, the protective structure should have the functions of retraction and redeployment.

From the above analysis, it can be seen that the advantages of active inflatable protection system over the traditional protective structure are as follows:

(1) The system can be deployed quickly and flexibly, and is not restricted by the overall satellite layout, envelope size, rocket launch and other bottleneck factors.
(2) The flexible protection system does not need an additional support structure. The high-strength flexible protection layer made of woven ceramic fabric or aramid fiber has a large mass ratio, an enough after-deployment spacing, and a higher protection efficiency.
(3) The flexible housing does not need the comprehensive consideration of multiple protections, and thus has a lower engineering difficulty.
(4) The interaction between the gas medium and the debris cloud formed by the hypervelocity collision can further reduce the debris erosion.

### 5.5.7 Space Radiation Protection Structure

In addition to the above active and passive protection designs against space debris, the threat of space radiation environment to the on-orbit safety and reliability of spacecrafts should also be considered. The space radiation protection can be divided into local protection and whole-spacecraft protection. Local protection is to use mass shielding for radiation protection, usually requiring the uniform shielding distribution over various items. The layout of instruments, devices, fuel and other materials in the spacecraft cabin can be optimized in order to obtain an uniform mass-shielding thickness on sensitive devices in the spacecraft cabin or in the astronaut cabin. To prevent potential transient high-intensity space radiation (such as solar particles), the sensitive devices/components or the astronauts and their sensitive parts can be locally shielded. Shielding layers and shielding compartments can be added to the sensitive areas to shield the sensitive devices from radiation in a transient high-intensity environment.

Like space debris protection, the radiation protection is also divided into passive protection and active protection. Passive protection is to use various materials to shield the sensitive parts from radiation. Active protection is to use electrostatic field, plasma, magnetic field and other active means to prevent high-energy particles from entering the protected object. The whole-spacecraft protection is also based on active protection, using magnetic or electric field to guide the charged particles away from the spacecraft so as to achieve the radiation protection. Generally, high energy protons are handled with an artificial magnetic field, and the electrons are handled with an artificial electric field. The active protection techniques for dealing with high-energy charged particles in space mainly fall into three categories: electrostatic field technique, superconducting magnetic field technique, and the technique of electric and magnetic fields synergy.

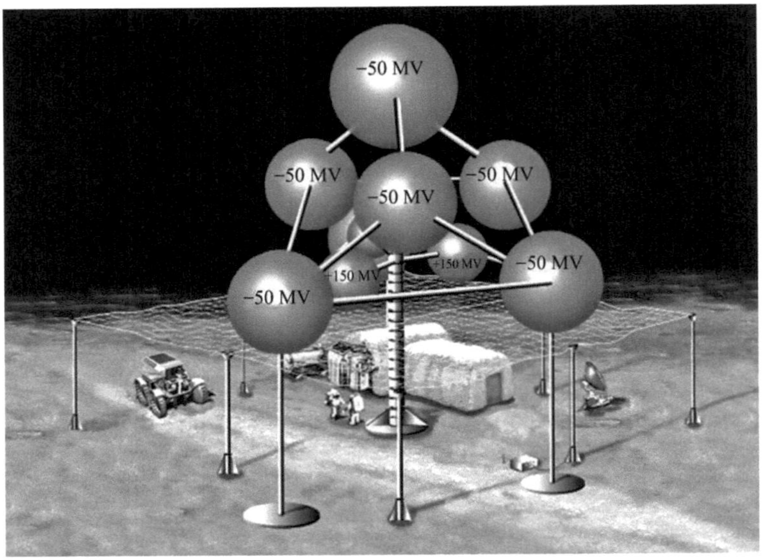


**Fig. 5.15** Electrostatic protection scheme developed by ASRC Aerospace Corporation (see the colored illustration)

The ASRC Aerospace Corporation explored an active protection technique for handling the high-energy charged particles in lunar base, as shown in Fig. 5.15. Multiple high-voltage electrostatic spheres are combined in a reasonable way into a protective electric field to stop the incident high-energy protons. This design can basically block the way of high-energy protons below a certain energy level.

The Beijing Institute of Spacecraft Environment Engineering has initially studied the active radiation protection based on superconducting coils. They use the superconducting coils to generate a strong magnetic field, which can deflect the incident charged particles and keep them out of the spacecraft. The technique they propose is to place ring brackets around the spacecraft and generate four groups of current in the same direction in each bracket, so that a large magnetic field is formed in the middle of the coils, but a very small magnetic field is formed outside the coils. This device is similar to the superconducting Tokamak magnetic-confinement device.

## 5.6 Basic Principles and Improvement of Spacecraft Protection Design

### *5.6.1 Principles of Spacecraft Protection*

For a spacecraft (especially a manned spacecraft), the design of space debris protection and space radiation protection must be considered in the stage of scheme design. At first, the protection index is determined according to the mission type and spacecraft requirements. Then, the preliminary risk analysis and protection design are carried out, followed by the optimization of the design scheme. Finally, the scheme that meets the protection index and requires the smallest protection weight is given.

The general debris protection measures to be taken for a spacecraft are prioritized as follows:

(1) Adjust the spacecraft attitude to reduce the collision risk. Orient the stronger part of the spacecraft towards the direction of larger impact flux (e.g. the direction of spacecraft movement), and orient the weaker part towards the direction of smaller impact flux (e.g. the direction opposite to the direction of spacecraft movement, or the direction towards the ground).
(2) Rationally design the spacecraft layout to reduce the collision risk. For example, place the parts with stronger impact resistance in the higher risk area and the parts with weaker impact resistance in the lower risk area.
(3) Reduce the collision risk through redundancy design, such as the redundancy design of external cable bus.
(4) Increase the thickness of the main structure of the spacecraft to reduce the collision risk.
(5) Add a special protective structure to reduce the collision risk.

The space radiation protection for a spacecraft is mainly divided into the astronaut protection and the protection of sensitive devices and components, which are generally realized through local protection. For example, the circuit boards and semiconductors susceptible to static electricity are protected against static electricity through electromagnetic compatibility design, electrostatic protection coating and shield housing adding. Mass shielding is used for the radiation protection of the spacecraft. The layout of instruments, devices, fuel and other materials in the cabin can be optimized in order to obtain an uniform mass-shielding thickness in the astronaut cabin or on sensitive devices in the spacecraft cabin. In addition, to prevent potential transient high-intensity space radiation (such as solar particles), an emergency shielding compartment should be set up for astronauts to ensure their safety and health in a transient high-intensity environment.

### 5.6.2 Design and Improvement of Protective Structures

Among the passive protection structures of a spacecraft, the rear wall is required to have high resistance to debris penetration and impulse load impact. However, the rear wall is a part or component/subsystem of the spacecraft cabin, so its protection scheme is mainly determined by the overall design requirements of the spacecraft, which leave little room for the rear wall to improve. The shield spacing should be moderate. If the spacing is too small, the protective capacity cannot be given full play. If the spacing is too large, the space will be limited, the mass of additional structures will be increased, and the protective capacity can't be significantly improved. A shield with excellent protection performance should have a strong ability to break, melt or vaporize the projectile and have a high strength-to-mass ratio, while ensuring a low damage threat of its debris to the rear wall and a large diffusion angle of debris cloud[3] From the above analysis, it can be seen that the shield has the greatest potential for improvement. According to the above analysis of various protective structures, the following measures can be taken to improve the protective performance:

(1) Add a mesh shield in front of the continuous bumper shield to effectively improve the effect of projectile breaking and increase the diffusion area of debris cloud. In addition, the secondary debris generated by the mesh itself in a collision will become smaller so as to reduce the damage to the rear wall.
(2) Add a foamed-aluminum shield. The perfect ballistic characteristics of such a shield can greatly improve the melting degree and diffusion angle of projectile fragments.
(3) Choose a high-fiber protective material. The selection of a fiber material with high elastic modulus and large strength-to-mass ratio can effectively improve the performance of the protective structure. In addition, the secondary debris of this material generated by the projectile impact is small with little damage to the rear wall.
(4) Match the rear wall material with the protective structure. Where permitted, the rear wall material should be appropriately adapted to the chosen protective structure. For example, the main failure form of the rear wall with a small-spacing Whipple structure is spalling. In this case, an anti-spalling lining can be added to the rear wall, or an anti-spalling material such as honeycomb board or laminated board can be chosen as the rear wall material.
(5) Process the back wall. The processing of the rear wall texture can prevent the occurrence of spalling to a certain extent.
(6) Appropriately increase the total shield spacing of the protective structure. In the high-speed projectile collisions, the increase of the total spacing will lead to not only the increase of the protection capacity but also the increase of the structure mass. Therefore, the optimization technology of the total spacing should be used for comprehensive design and consideration.

---

[3] Yuan Jungang, Qu Guangji, Yan Jun. Progress in the Design Technology of Space Debris Protection Structures [C]. 2007 National Symposium on Structural Dynamics.

In the design of active protection mechanism, the mechanism is required to have the abilities of rapid deployment and retraction/redeployment. Therefore, the selection of flexible shield material, the design of folding mechanism and the design of rapid inflation mechanism need to be considered comprehensively. First of all, the type of flexible gasbag fabric determines whether the inflatable structure can expand rapidly without rupture and can withstand the huge pressure generated during rapid inflation and deployment, and whether the projectile can be broken into pieces without exploding at the moment of debris incidence. The shield layer material is required to have enough strength and be easy to fold and stitch. In addition, the folding/deployment mechanism of the flexible shield should be specially designed to ensure that the inflatable structure can be quickly and reliably deployed according to the engineering requirements and can be reliably retracted. Finally, the design of the inflation mechanism directly determines whether the active protection system can be rapidly deployed. The gas quality, gas source, output pressure and other factors of the inflatable structure should be comprehensively considered to ensure rapid inflation, a small size, good stability and reliability, and no gas leakage and explosion.

## References


1. Gong, Z., Zhao, Q., Li, M., et al. (2019). Frontiers and prospects of debris protection research. *Space Debris Research, 19*(3), 2–13.
2. Shen, Z., Xia, Y., Yang, Y., et al. Spatial radiation protection materials and structures for spacecrafts. *Aerospace Materials and Technology, 20*, 50(2), 1–7.
3. Yu, H. (2007). *Research on environment risk assessment techniques for spacecrafts and micrometeoroids*. Harbin Institute of Technology.
4. Wang, B., Xu, Y., & Wu, Q. (1990). Interaction between spacecraft and space plasma environment. *Acta Astronautica, 3*, 10–15.
5. Shigui, Z., & Jun, Y. (2014). Development of debris protection requirements and materials. *International Space, 6*, 49–53.
6. Ma, Z. (2017) *Evaluation of spacecraft survivability in space debris environment*. Harbin Institute of Technology.

# Chapter 6
# Design of Space Debris Protection System

## 6.1 Estimation of Spacecraft Collision Risk

In the region near LEO, the maximum speed of a spacecraft colliding with space debris can reach 15 km/s, and the average collision speed is about 10 km/s. The energy generated is about 500 kJ, which is equivalent to the impact energy of a 10 g aluminum ball at a speed of 10 km/s, and comparable to the impact energy of a 1.3 t car at a speed of 100 km/h on the ground [1]. The spacecraft collision with space debris happens from time to time. As of 2019, the ISS has made 25 maneuvers to avoid the debris impacts. More than 16 satellites have been publicly reported disabled or abnormal due to debris impacts. In recent years, the NASA's satellites have made more than 20 maneuvers a year to avoid space debris, a threefold increase from 5 maneuvers in 2008. In 2018, the ESA satellites made 17 avoidance maneuvers (including 15 maneuvers in LEO and 2 maneuvers in GEO)[1]. In 2009, a total of 58 craters with diameters greater than 0.25 mm were found in the aluminum alloy plates 01–04 B and 02–04 B, which protected the airlock module on the International Space Station (ISS) for nine years, as shown in Fig. 6.1.

With the help of the space shuttle, the NASA checked and analyzed the impact damage to the ISS modules, including multipurpose laboratory module (MLM), P6 truss battery module, fuel tank assembly (ATA) and airlock module shield. The glass porthole and radiator of the space shuttle were also checked. As a result, 75 craters with diameters within 0.1–1.5 mm were detected on MLM. 49 craters with diameters within 0.1–1 mm were formed on ATA during the 7-year orbiting. The STS-132 mission detected 15 craters (maximum diameter: 1.3 mm, maximum depth: 0.12 mm) on the shuttle's glass porthole and 12 craters on the radiator, none of which were perforated. Figure 6.2 shows the detected impact damage to some parts of the ISS.

[1] Marchal Ph. 2018 Space debris activities in France: Highlights [R]. 37th IADC Meeting, 2019.



J. Zhang et al., *Space Debris Research*, https://doi.org/10.1007/978-981-96-9570-6_6

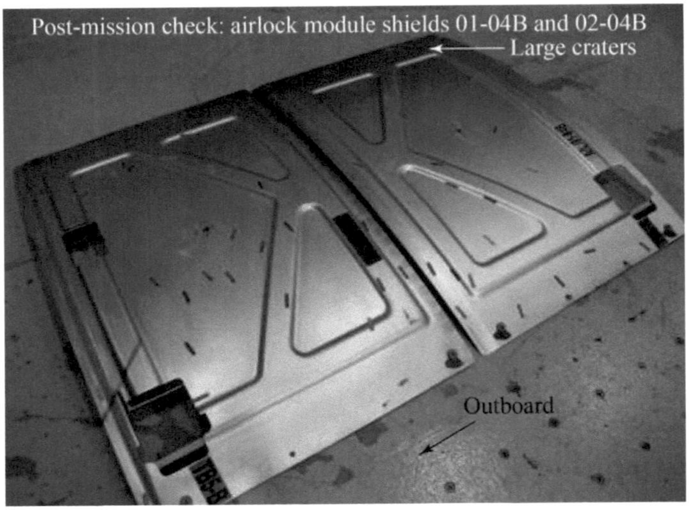


**Fig. 6.1** Recovered airlock module shield (see the colored illustration)

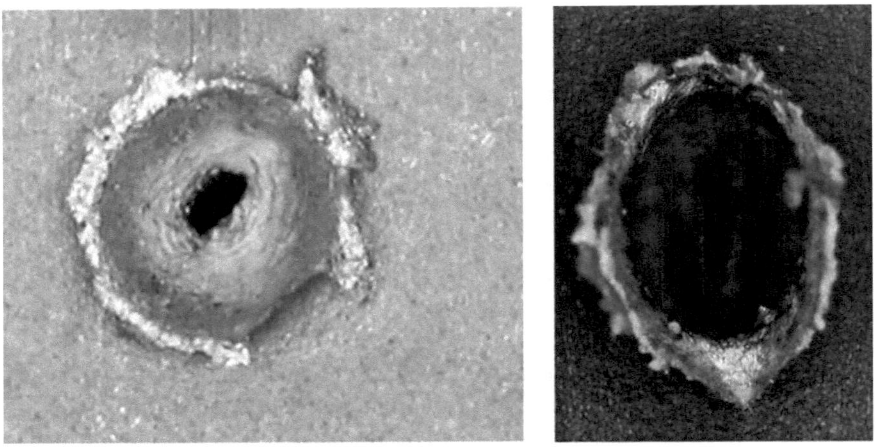

(a) (b)

**Fig. 6.2** Impact damage to some parts of the ISS (see the colored illustration) **a** Perforation of MLM A3 panel; **b** ATA impact damage

The assessment of spacecraft impact risk is an important basis for protection design. The Russian Central Research Institute for Machinery Manufacturing, part of the Russian Federal Space Agency (ROSCOSMOS), has upgraded the COLLO risk assessment software by integrating it with the LEO debris model and adding the capability of assessing the damage of secondary debris to spacecrafts. The COLLO software has also been cross-checked against the risk assessment software from other countries and organizations. The US is gradually refining the assessment of M/OD collision risk to the Space Shuttle and the ISS. The JAXA developed the M/OD risk assessment software "TURANDOT" and used it for analysis and calculation.

## 6.2 Ground Test Method for Hypervelocity Collision

Since the successful launch of an man-made Earth satellite in the 1950s, the safety of spacecraft operation in outer space, especially the safety in a micrometeoroid impact, has become a growing concern for spacecraft designers. To meet the objective protection needs of space objects (for example, space stations, space shuttles, manned spacecrafts and satellites), the space powers around the world have started to comprehensively construct and develop the ground-based hypervelocity collision simulation test equipment, including the launch device and velocity measurement device, in order to understand the impact damage effect on spacecrafts and design more effective protection structures. Through a large number of ground hypervelocity collision tests, the cratering, perforation, ballistic limit velocity, secondary debris cloud and other characteristics and effects of different aerospace materials and shields under the impact of the projectiles (particles) with different speeds, sizes, shapes and materials have been studied, verified, evaluated and obtained, providing the supporting data for the development of high-performance advanced protection materials and the shield design. The typical characteristic of hypervelocity collision is that the pressure generated by the collision is much greater than the initial strength of the projectile and target plate. According to the thickness of target plates, the hypervelocity collision problems can be divided into three categories, namely thin plate collision, medium plate collision and thick plate collision. For the target plates of different thickness, since the compression waves generated by collision and the sparse waves reflected from free surface have different action processes and products in the projectile and target plate, the final test results are also different. For thin plate, a debris cloud is mainly formed behind the target plate.

The existing ultra-high speed (UHS) launchers developed at home and abroad mainly include light gas gun, electric gun, energy-gathered accelerator, plasma accelerator, and laser-driven flyer. Among them, light gas gun, electric gun and energy-gathered accelerator are mainly used in the hypervelocity impact test of millimeter-sized space debris, and plasma accelerator and laser-driven flyer are mainly used in the impact test of micrometer-sized space debris. The launching capacities of various launchers are shown in Fig. 6.3.

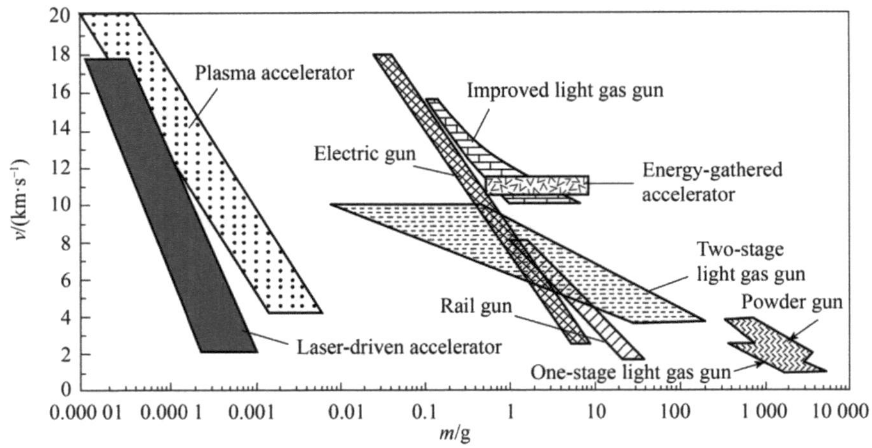


**Fig. 6.3** Applications of various launchers

## 6.2.1 Light Gases Gun Technology

Light gas gun (LGG) is the most developed and widely used ground-based high speed/ UHS simulation launch facilities. In particular, two-stage LGG is the most widely used and was first made by the New Mexico Institute of Mining and Technology in 1948. A typical two-stage LGG is mainly composed of a compression stage and a firing stage. It works in the following way. The energy-stored gas in the first stage (compression stage) pushes the piston to compress the light gas (such as helium, hydrogen, etc.) pre-filled in the second stage (firing stage), which, in turn, pushes the projectile to accelerate so as to realize HS/UHS firing.

A complete set of two-stage LGG hypervelocity collision test equipment mainly includes a launching system, a velocity measurement system and a target chamber. To achieve different test purposes, a pressure sensing system and a measurement system can be added. The function of the launch system is to fire a projectile with the required speed to hit the target. A metal projectile should be fired with the aid of a non-metallic sabot. In the launch system, the gas used for the first-stage drive is usually high-pressure nitrogen gas or the high-pressure gas produced by powder explosion, and the gas used for the second-stage drive is hydrogen gas. The velocity measurement system is used to measure the velocity of the projectile before it hits the target.

The two-stage LGG can directly fire non-metallic projectiles, but the direct firing of metal projectiles will seriously damage the launching tube. Therefore, a metal projectile should be fired with the aid of a non-metallic sabot. During firing, the sabot is used to center, support and protect the projectile to ensure that it does not contact with the launching tube and high-temperature driving gas. When flying out of the muzzle, the projectile is separated from the sabot and flies along the original ballistic trajectory. Meanwhile, the sabot is blocked or captured by a preset device. This technology is called projectile-sabot separation. The techniques of projectile-sabot separation are usually classified into pneumatic separation and mechanical forced separation. In the technique of pneumatic separation, the sabot is composed of two or four pieces of polycarbonate, and its outer diameter is the same as the diameter of the launching tube. During firing, the sabot flies together with the projectile. When flying out of the muzzle, the projectile is separated from the sabot under the joint action of the shock wave at the outlet of the launching tube and the surrounding flow field. Mechanical forced separation is to use a stop mechanism on the trajectory to forcibly separate the projectile from the sabot. This method can be realized in a vacuum target chamber, thus avoiding the loss of projectile velocity due to the presence of gas in the target chamber.

According to the diameters of the launching tubes, the two-stage LGGs can be divided into three types: small-calibre LGG, medium-calibre LGG and heavy-calibre LGG. Among them, the small-calibre LGG has a launching tube diameter of 4–5 mm and a pump tube diameter of 15 mm, and is suitable for firing the projectiles (or particles) with a diameter of 2–100 mm to a speed up to 8.5 km/s. The medium-calibre LGG has a launching tube diameter of 6.5–15 mm and a pump tube diameter of 40 mm, and can accelerate the projectile with a mass of 5 mg to a maximum speed of 10 km/s. The heavy-calibre LGG has a launching tube diameter of 20–50 mm and a pump tube diameter of 65 mm or 100 mm, and can accelerate the projectile with a mass of 1.0 g to a maximum speed of 8 km/s.

Theoretically, the speed of a LGG can reach 11 km/s. However, due to the limitation of the structural strength and ablation resistance of the gun body, the ultimate launching speed of most of the two-stage LGGs is below 8 km/s, which is still far behind the average relative speed of space debris (10 km/s). In order to increase the launching speed, some new launch techniques have been developed, such as three-stage LGG, electric gun, and plasma accelerator. The three-stage LGG usually has a preheating stage added to two-stage LGG, which is mainly composed of a compressed air chamber, a large-mass piston, a preheating pump tube and a link port. It can further increase the firing speed of the projectiles. In the 1960s, the McGill University built a three-stage LGG that could fire a 1.1 g polycarbonate flyer with a diameter of 12.7 mm to a speed of 10.5 km/s. The three-stage LGGs of San Diego National Laboratory and Livermore National Laboratory could fire 3.1 g projectiles to a speed of 10.1 km/s. Based on two-stage LGG technology, the EMI in Germany developed a new two-stage LGG with a single powder chamber and double pump tubes, as shown in Fig. 6.4. The gun could effectively reduce the dynamic load borne by the parts of the gun body when a projectile was loaded, thereby increasing the firing speed and keeping the gun in good performance and stable and reliable condition. In January 2011, the gun underwent the first test, in which the firing speed of a 100 mg projectile reached 6.4 km/s.

**Fig. 6.4** New LGG (see the colored illustration)

### *6.2.2 Electromagnetic Gun Technology*

Electric gun is an UHS launcher which uses electric energy as power to accelerate the projectiles. As a kind of electric gun, the electromagnetic gun uses electromagnetic effect to accelerate the projectiles. The electromagnetic guns are classified into three types by structure: coil gun, rail gun and re-connection gun. Compared with traditional gunpowder gas, the electromagnetic gun accelerates the projectiles over a longer period to a higher speed, for example, accelerating the 4.3 mg Kapton flyer to 18 km/s as demonstrated in a test.

Coil gun is an UHS launcher that uses the interaction between the induced magnetic fields of energized coils to accelerate the projectiles. As the earliest development form of electromagnetic gun, the coil gun is mainly composed of acceleration coils and projectile coils. The acceleration coils are fixed on the gun barrel. When an AC current is applied, an alternating magnetic field will be generated, leading to the generation of an induced current in the projectile coils. The magnetic field generated by the acceleration coil current interacts with the magnetic field generated by the induced current to produce a Lorentz force, thus driving the projectile out and accelerating its speed. When a number of acceleration coils fixed on the barrel are provided with the AC current in turn, the projectile can be driven step by step by Lorentz force to achieve a very high firing velocity.

The rail gun consists of two parallel metal rails, an armature sliding along the axis of guide rails, and a high-power pulse generator. The guide rails are connected with the pulse generation network. The armature is located between the two rails and is used for switching and short circuiting. The projectile is placed on the rails in front of the armature to form a closed circuit. The metal projectile needs to be wrapped with an insulating material or supported by a sabot. When the projectile is fired, the pulse generation network will supply power to one guide rail, and the power will flow through the armature to the other guide rail. When a strong current flows through the two parallel rails, two strong magnetic fields in opposite directions will be generated between the two rails and will interact with the magnetic field generated by the armature to produce a strong electromagnetic force. The electromagnetic force will drive the armature and the projectile in front of it to move at an accelerated speed along the guide rails, so that the projectile can be fired at an ultra-high speed. Usually, the electromagnetic force applied on the rails lasts for only a few milliseconds. When the projectile leaves the rails, the remaining energy will be either channeled to the pulse generation network through the muzzle shunt, or formed into an arc and then dispersed in the air.

Re-connection gun is a multistage-acceleration non-contact electromagnetic launcher, which has two (upper and lower) rectangular coils with a certain gap between them. When passing through the gap, the projectile is accelerated by the strong magnetic field force generated by the two coils. This gun can not only fire large-mass projectiles as done by an AC coaxial coil gun and fire UHS projectiles as done by a rail gun, but also provide the projectiles with a higher peak acceleration force that is not much different from the average acceleration force so that the projectiles can obtain more uniform acceleration. The main advantages of re-connection gun include: no contact between the projectile and the coil, and relatively small ohm loss; higher energy transferred to the projectile per unit length than in other electromagnetic launch methods; high flight stability of the projectile, and no vibration in the directions of roll, yaw and pitch; high acceleration efficiency. The re-connection guns can be classified into: plate-projectile type and columnar-projectile type according to the projectile shape; or the type without capacitor and the type with capacitor depending on whether the gun is connected with an excitation pulse capacitor; or box-coil type, square-solid-coil type and round-coil type depending on whether the coil is wound; or single-stage type and multistage type according to the number of stages.

### 6.2.3 Other Launch Techniques

Other launch techniques mainly include electric heating gun, dust electrostatic accelerator and shaped charge launcher. Electric heating gun is an UHS launcher which can completely or partially make use of an electric heating medium to accelerate the projectiles. The electric heating guns can be classified into: direct-heating type and indirect-heating type according to the modes of plasma generation and operation; or back-injection type and side-injection type according to the mode of plasma injection; or chamber discharge type and tube discharge type according to the category discharge heating mechanism.

Dust electrostatic accelerator is a test facility for the UHS firing of dust particles with sizes below micron-level based on the theory of Van de Graaff DC high-voltage electrostatic particle acceleration. It is mainly composed of a Van de Graaff accelerator, a transmission tube and a target chamber, with a structure shown in Fig. 6.5.

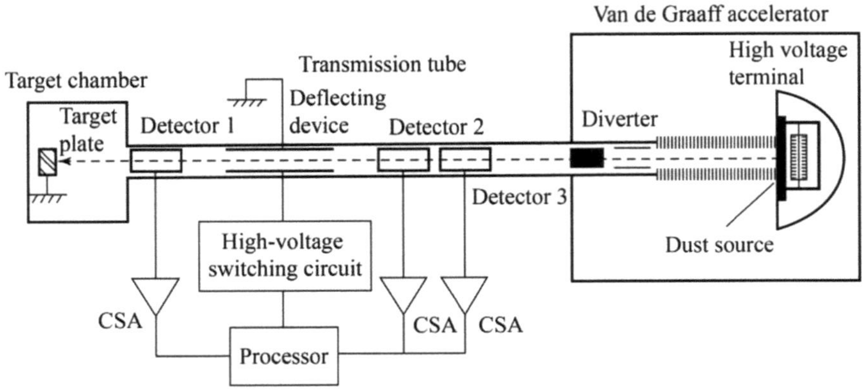


**Fig. 6.5** Structure of dust electrostatic accelerator

Shaped charge launcher is an UHS launching facility which uses the detonation pressure of high energy explosive to converge a conical metal charge liner (such as copper liner or aluminum liner) along its axis into a metal jet. The launcher works in the following way. After the shaped charge is detonated, the metal charge liner is crushed under the detonation pressure of the charge and converges along its axis to form a high-speed metal jet. The high-speed metal jet particles with the required length, as well as the slug, are captured by using a jet particle catcher and a slug catcher and controlling the thickness of the PVC plate.

### *6.2.4 Relevant Ground Impact Tests*

The ESA simulated the failure mode of a new European solar array on the ground using a two-stage LGG, analyzed the impact failure threshold, and studied the degree of impact-induced plasma generation and the influence of plasma cloud expansion on the array performance. In addition, the ESA studied experimentally the UHS impact damage characteristics of vulnerable spacecraft components such as cable harness, electronic unit and fluid pipeline. In the test, the cable harness was powered; the electronic unit was working, containing a processing unit and a memory chip; the fluid pipeline was under pressure.

The University of Padua, Italy, affiliated to ASI, studied the anti-sputter UHS frontal/oblique impact of space debris on solar silicon cells, aluminum alloy panels, and multiple insulation layers of satellite surface in a ground acceleration simulation test, used Autodyn software for numerical simulation, and finally obtained the distribution characteristics of anti-sputter debris. In all conditions of the test series, three types of target plate structure were adopted, including an Al-6082-T6 plate with a thickness of 3–10 mm, an Al-6082-T6 composite target plate with a thickness of 3 mm covered with thermal insulation material, and a solar silicon cell unit. The test

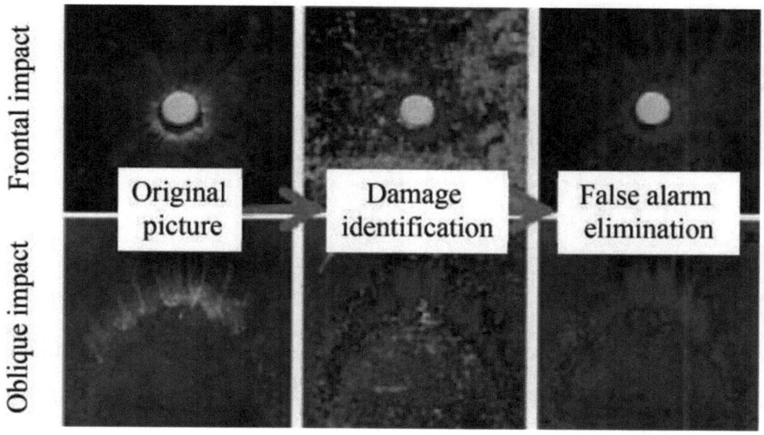


**Fig. 6.6** Craters on the observation plate (see the colored illustration)

projectiles had three diameters (1.0, 1.5 and 2.3 mm), four impact angles (0°, 45°, 60° and 80°), and two impact velocities (4.0–4.5 and 5.0–5.5 km/s). The observation plate used in the test was a 2 mm-thick copper plate. Figure 6.6 shows the distribution of impact craters on the observation plate obtained in the partial frontal impact test and oblique impact test.

In addition, the EMI in Germany preliminarily studied the distribution law of the secondary debris generated in a collision between a large piece of debris and a whole satellite through testing and numerical simulation,[2] and initially quantified the size and location distribution of the secondary debris. In the test, the impact projectile was an aluminum column with both the length and cross-section diameter of 12 mm and an impact velocity of 1.9 km/s. The test target was a 10 cm × 10 cm × 10 cm aluminum frame structure referring to the CubeSat satellite. The projectile collided with the target in three ways: edge impact, corner impact and side impact. At the same time, the SOPHIA software compiled by EMI was used for simulation calculation in the impact test of the simulated satellite. The simulation result was compared with the result of actual impact test, as shown in Fig. 6.7.

The University of Dayton in the United States conducted an experimental study on the characteristics of the debris clouds formed by the UHS impact of the projectiles with different shapes, as shown in Fig. 6.8. The results show that the shape and impact angle of a projectile have important effects on its fragmentation and the distribution characteristics of the formed debris cloud. A non-spherical projectile is more likely to produce one or more large fragments at the target edge, turning the debris cloud into a serious threat to the rear plate. The inclined cylindrical projectile has a stronger ability to damage the rear plate. This destructive power is strongest when the angle of inclination is about 45°. Inside the debris cloud formed by a disc-shaped projectile

[2] Yuan Jungang, Qu Guangji, Yan Jun. Progress in the Design Technology of Space Debris Protection Structures [C]. 2007 National Symposium on Structural Dynamics.

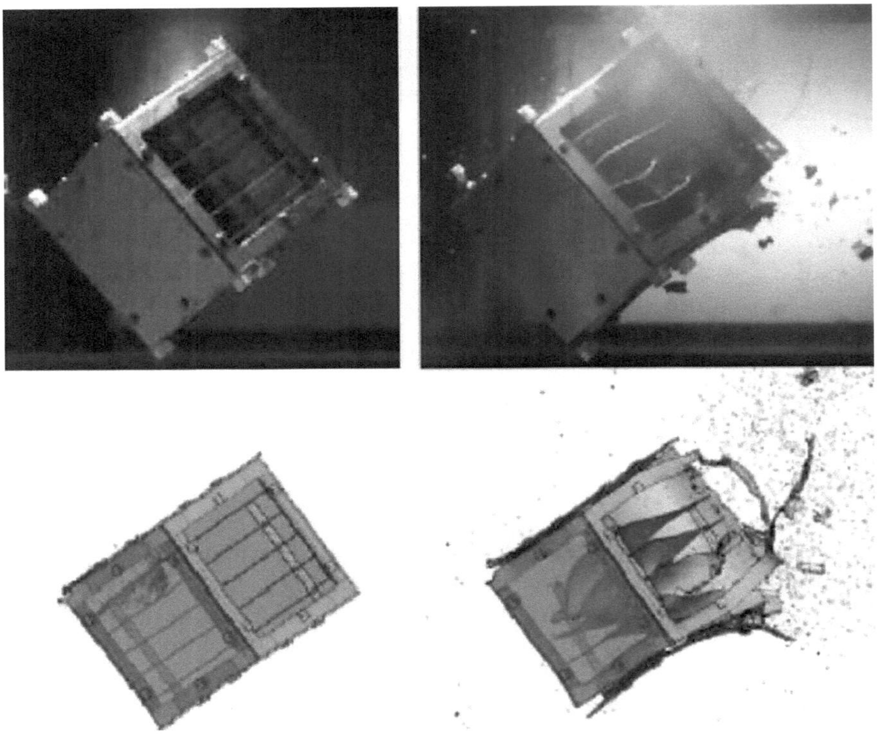

**Fig. 6.7** Comparison of ground impact test result and numerical simulation result (see the colored illustration)

is a columnar structure composed of both the projectile and the shield material. With the movement of the debris cloud, the middle columnar structure exhibits almost no diffusion and a small shape change, and thus is not sensitive to the protective spacing. The rear plate is mainly damaged by the middle column of the debris cloud.

In 2003, the Protection Analysis Department of NASA Johnson Space Center studied a variety of risk sources including space debris impact that might have led to the failure and crash of the space shuttle "Columbia" in that year. According to their investigation report, the existing risk assessment model lacked the ability to evaluate the damage from the impact of non-spherical projectiles, which led to the existence of an uncertainty risk in the orbiting spacecraft. Finally, the department urged the establishment of the parameters related to projectile shape in the future debris risk assessment model in order to predict the damage characteristics of shape effect.

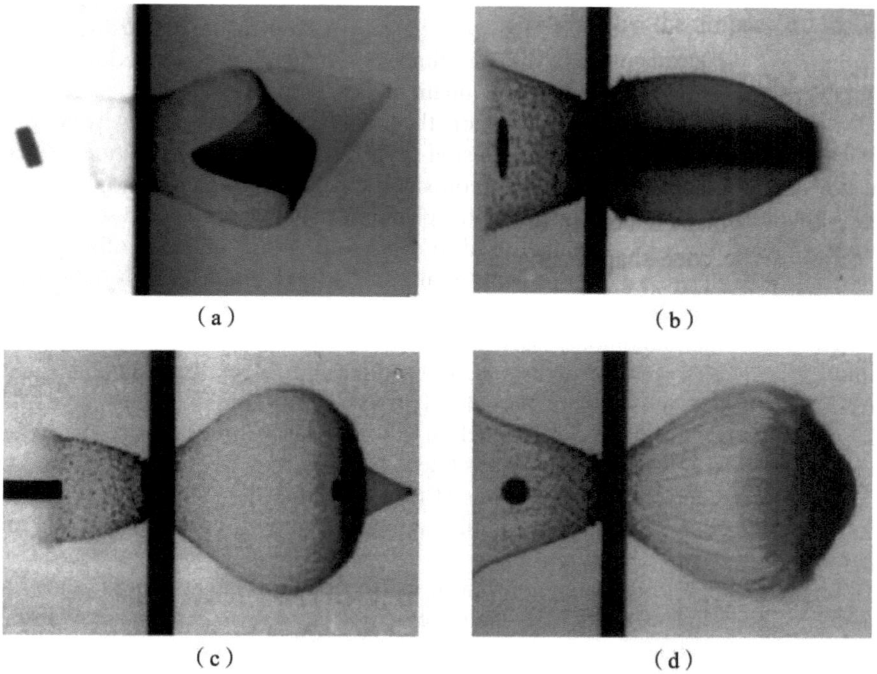


**Fig. 6.8** Shapes of debris clouds formed by the UHS impact of the projectiles with different shapes **a** Short cylinder; **b** disk; **c** long cylinder; **d** sphere

## 6.3 Numerical Simulation Method for Hypervelocity Impact

Compared with low-speed collision, the UHS collision has short impact time, large action range, strong nonlinearity and complex damage effect. Meanwhile, high temperature and high pressure, high frequency oscillation, very large deformation and even phase transition will occur in the impacted area. Therefore, the theoretical research on UHS collision is extremely challenging. In addition, the capacity of the hypervelocity impact equipment is limited, and the test cost is very high. Thus, it is unrealistic to support the study of hypervelocity impact solely by impact test. In this context, numerical simulation has become an important means for supporting the development of hypervelocity impact research.

The research on the numerical simulation of hypervelocity impact began in the 1950s. Since the impact pressure generated by the hypervelocity impact was much greater than the strength of the projectile and target that were similar to incompressible fluids during the impact process, the fluid model was mainly adopted in the early stage and then the fluid elastoplasticity model considering the influence of fracture and other factors was introduced in the later stage. From the perspective of spatial dimension, the simulation model has evolved from one dimension to two dimensions and then to three dimensions. The one-dimension model has a simple logic and a small program volume, and can obtain qualitative estimation of the physical process. However, due to excessive simplification, the model can only quantitatively approximate the test process. The two-dimension model is the main analytical means, and can accurately describe the head-on collision between a symmetric projectile and the structure and simulate many complex phenomena in experimental research. The three-dimension model is mainly used to deal with various oblique collisions. Its theoretical calculation is not limited by any factor, and can accurately reflect the actual object shape and physical process.

The algorithms commonly used in the numerical simulation of hypervelocity impacts include: Lagrange algorithm, Euler algorithm, Arbitrary Lagrange-Euler (ALE) algorithm and Smooth Particle Hydrodynamics (SPH) algorithm. In the Lagrange algorithm, the mesh is fixed on the material and moves with the material. In the Euler algorithm, the mesh is fixed, while the material moves in the mesh. The ALE overcomes the shortcomings of Lagrange algorithm and Euler algorithm to a certain extent. The SPH algorithm is a meshless algorithm widely used in the simulation of hypervelocity impacts.

Lagrange algorithm is an algorithm in which the mesh is fixed on the material and the mesh points coincide with the material points, so there is no relative motion between mesh points and material points. In addition, there is no convection term in the control method, which simplifies the governing equation and the solving process. This method can track the motion of objects easily and can provide help for dealing with the interface and boundary between materials. However, when the object has a large deformation, the mesh will be distorted, resulting in the generation of large errors and the failure to simulate the formation of debris cloud and do the relevant calculations.

The Euler algorithm uses a fixed coordinate system in which the mesh is fixed in space and remains unchanged during the deformation of the object. Therefore, this method has no large mesh distortion and is very suitable for solving the large-deformation problems. In addition, the Euler algorithm can deal with the material diffusion and mixing, allowing several materials to exist in a spatial mesh. Because the Euler algorithm uses a fixed mesh, the mesh is required to expand to the area where the object is likely to move. This will lead to the adding of more nodes and the increase of computation time, especially in the simulation of hypervelocity impacts.

Smooth Particle Hydrodynamics (SPH) algorithm is a meshless Lagrangian algorithm, in which the computational domain of a problem is characterized by filling a series of meshless SPH particles. In the calculation, there is no need to use a predefined mesh to provide interconnection information for SPH particles. The meshless particles in SPH algorithm not only serve as interpolation approximation points, but also carry the material properties. They can move under the internal and external forces and thus have higher flexibility. The construction of a conservation equation is not affected by the random distribution of SPH particles, and can naturally deal with nonlinear dynamics problems in the materials with very large deformation, which reflects good adaptability of this algorithm. Just because of the harmonious combination of meshless particle characteristics, Lagrangian algorithm characteristics and adaptive characteristics, the SPH algorithm is increasingly widely used in the numerical simulation of nonlinear dynamics of the solid materials with large deformation.

In terms of physical model, the material will undergo phase changes such as melting and vaporization under the condition of an UHS impact above 10 km/s. To accurately describe the phase changes during the impact process, an equation of complete multi-phase state that can describe the three phases (solid, liquid and gas) must be available. The investigation and analysis show that the Tillotson, SESAME, ANEOS and GARY equations are the main equations of three-phase state used in the numerical simulation of hypervelocity impacts above 10 km/s.

Tillotson equation is a simplified equation of state formed by combining Hugoniot line with Thomas-Fermi model, and is the analytical equation of state most widely used in collision research. The equation has two complex forms for describing the compressed state and expanded state respectively while covering the solid, liquid and gas phases and corresponding to a series of parameters separately. The main advantage of Tillotson equation is that it is applicable to a high pressure region, with an error of about 10% under 100 TPa and about 5% under 500 GPa. However, the equation of state cannot provide information for the calculation of material temperature or entropy, and cannot describe the changes of pressure and density in the two-phase mixing region. Therefore, the melting and vaporization of materials cannot be described effectively. In addition, the Tillotson equation is not monotonous in the low-energy expansion zone, where an unreasonable negative pressure occurs.

The ANEOS equation of state uses a complex computational code derived from physical approximations corresponding to different phase regions, compiled by FORTRAN, and invoked by a hydrodynamics computational program. The pressure, temperature and density are derived from the Helmholtz free energy to ensure thermodynamic consistency. One of the main advantages of ANEOS over the analytic equation of state is that it can describe (finitely) important processes in a phase-change impact. The main disadvantage of ANEOS is that the gas is treated as a monatomic molecule so that the gas-liquid critical point is overestimated. A processing method for diatomic molecular gas has been introduced by Melosh into ANEOS to make the description of gaseous phase more realistic. This update reflects the development of ANEOS towards a more realistic equation of state.

SESAME database is a tabular state-equation database developed by LANL National Laboratory, consisting of the equations of complete state. Its thermodynamic data mainly include internal energy, pressure and Helmholtz free energy, which are all the functions of temperature and density. Given the temperature and density of a material, the thermodynamic state data (such as pressure and internal energy) of the material can be obtained.

GARY equation is a very practical three-phase equation of state established by E. B. Royce based on Grover's equation of liquid state and Young-Alder's equation of gaseous state. In this model, different phase regions (solid, liquid and gas) are described by different equations of state. For example, solid phase is described by Gruneisen equation of state, and liquid phase is obtained by correcting the solid phase equation. GRAY equation is an equation of incomplete state, that is, it cannot directly give the temperature.

In terms of constitutive model, since the strength of materials far from the collision area should be considered in a hypervelocity collision, the current development of CTH program generally adopts the ideal elastoplastic model, and sometimes Johnson-Cook model and Steinberg-Cochran-Guinan model.

## 6.4 Spacecraft Vulnerability Analysis and Disintegration Model

The vulnerability research is mainly to: (1) obtain the damage thresholds of the spacecraft components/subsystems after a debris impact, and the relationship between their damage degree and impact parameters; (2) obtain the damage mode and mechanism of the spacecraft components/subsystems; and (3) establish the damage criterion. This research is very challenging because the spacecraft has a large number of components/subsystems, their failure modes and their influences on the system are very different, and their sensitivities to debris impact are also different. Vulnerability will be one of the main research topics of the IADC Protection Working Group over the next few years. The IADC has deployed the research on spacecraft components/ subsystems such as solar arrays, cables, batteries, electronic boxes, multilayer insulation materials, pressure vessels, fluid pipelines, transparent materials, and satellite structure materials.[3] The United States has developed the model HIVAM, which can quickly evaluate the damaged area and damage probability of satellite components after a debris impact. By using the output of the vulnerability analysis model COVART and FASTGEN code for aircraft and missiles, the model is directly coupled with the low-speed impact model to ensure the smooth transition of debris impact at low speed, high speed and ultra-high speed. The HIVAM model can quantitatively and rapidly calculate the vulnerability area and damage probability of satellite components in an UHS debris impact to support the design and vulnerability analysis of the satellite protection system.

The spacecraft breakup model under space debris impact is very important to establishing a high-precision space debris environment model, evaluating the space emergencies, and analyzing the debris situation evolution. Since 2014, the United States has begun the establishment of the DebriSat Project, a satellite breakup model based on hypervelocity impact, which is scheduled to be completed by 2020. The project was jointly funded by the NASA's ODPO and the USAF Space and Missile Systems Center (SMC). Led by the ODPO (specifically at the Johns Space Center or JSC), the Aerospace Corporation, the USAF Arnold Engineering Development Complex (AEDC) and the University of Florida took part in the project in order to improve the USAF Satellite Orbital Debris Characterization Impact Test (SOCIT), a previous satellite breakup model based on hypervelocity impact. A typical satellite

[3] Eric Christiansen, Martin Schimmerohn. IADC 37 WG3 Protection Status [R]. 37th IADC Meeting, 2019.

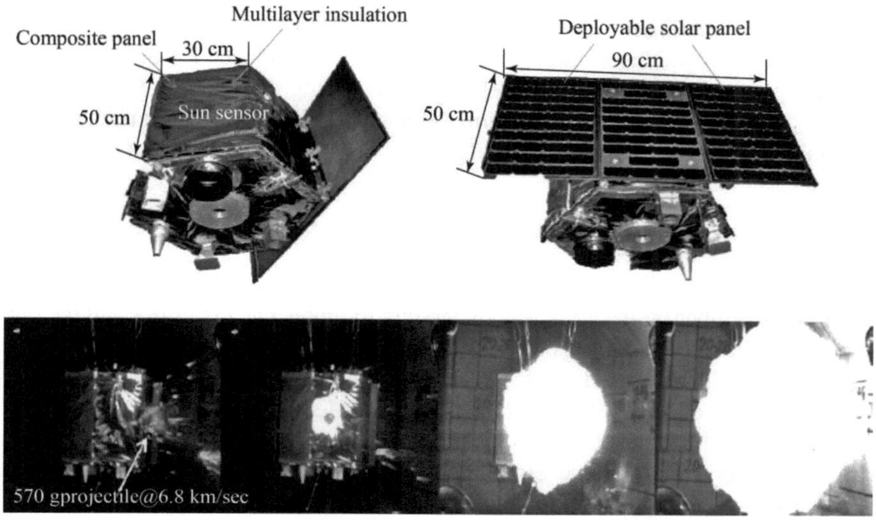


**Fig. 6.9** Test of post-impact satellite breakup model (see the colored illustration)

breakup test is shown in Fig. 6.9. In the experiment, a 45.7 cm × 45.7 cm × 50.8 cm satellite was simulated to obtain the impact energy threshold for satellite breakup and the distribution law of the debris generated by the impact, study the impact effect of non-spherical projectiles, and verify the accuracy of the numerical simulation.

## 6.5 Key Techniques of Protection System Design

Different from the protective shield special for a manned spacecraft, the non-manned spacecraft has very strict requirements for onboard mass. Moreover, the flight attitude and overall configuration of a non-manned spacecraft (e.g. satellite) are determined by the constraints such as satellite missions and equipment functions, without considering the impact of space debris in general. According to international experience, the satellite protection design is generally achieved by protective reinforced material or cladding composite material.[4] Therefore, the satellite protection depends more on a reliable structure, good performance of the protective material, and optimal design of the on-board protection layout. The key techniques for developing the satellite protection are summarized as follows:

1. **Technique of ground-based hypervelocity impact test**

The hypervelocity impact test provides a foundation for the whole passive protection technology [2], for example, by providing a material model and database for numerical simulation and a practical engineering model for risk assessment and structural

[4] Eric Christiansen, Martin Schimmerohn. IADC 37 WG3 Protection Status [R]. Rome: 37th IADC Meeting, 2019.

optimization, defining the damage characteristics of structures and materials and the failure modes of satellite components, and providing guidance for the selection of protective structural materials. The space debris hitting the satellite has an average velocity of 10 km/s and the maximum velocity of 15 km/s, so the ground launch technology needs to cover the velocity range of 0–10 km/s and even higher than 10 km/s.

2. **Technique of numerical simulation in hypervelocity impact test**

Due to the diversity of structures and materials, the complexity of space debris impact, the limited number of ground-based hypervelocity impact test facilities and the high experiment cost, the countries have been focusing on the numerical simulation of hypervelocity debris impact as a key technology with great difficulty while strengthening the study of hypervelocity impact test. This technology can not only predict and verify the impact test result and make up for the deficiency of test data, but also solve the hypervelocity impact simulation and protection design, which are difficult to achieve in a ground test, through extrapolation and numerical simulation [3].

3. **Advanced protective structures and materials**

According to the requirements of satellite protection, the selection and application of local or global protective reinforcing structures and materials for the satellite are the key points of satellite protection. Therefore, the development of advanced protective structures and materials is very important.

4. **Satellite vulnerability and failure mode analysis technique**

The space debris environment can cause several mechanical failure modes of the spacecraft (for example, breakup, perforation and surface damage), and may bring mechanical, electrical and thermal effects. Most of the past studies mainly focused on the perforation damage to the sealed capsule of a manned spacecraft, thus generating a number of impact limit equations and impact limit curves. However, these equations and curves cannot fully meet the needs of debris risk assessment and debris protection research. Various failure modes of a spacecraft under the debris impact, ranging from the breakup failure of the spacecraft to the failure of a certain part/subsystem, should also be included into the scope of systematic research. The study of the component/subsystem failure modes is a long-term task in the future space debris research [4].

5. **Debris collision risk assessment and protection integration**

The system mainly includes the risk assessment module, protection design/optimization module, and database module. After being formed, the system can not only provide an important tool or means for the risk assessment and protection design of the spacecrafts in our country, but also serve as a platform for technical communication and result display at home and abroad [4].

## 6.6 Development Trend of Protection Technology

So far, the ESA has developed a numerical simulation method for the local damage caused by hypervelocity impact between a small impactor and a large target, and verified its effectiveness by comparing it with ground test.[5] In terms of hypervelocity impact and spacecraft protection, the ESA pays high attention to the following areas: reduction of the perforated structural designs; material models and composites; and the design margin required to compensate for the damage of non-spherical impactors or charge elements (such as solar arrays). The future research will focus on how to improve understanding of the results of a hypervelocity collision between two larger objects, paying more attention to debris characteristics (such as size distribution) and the resulting velocity changes and distribution, and the directional changes in collision geometry and energy functions. The ground testing method is feasible, but has a large difficulty. Therefore, the main focus is still on numerical simulation method.

During the Eleventh Five-Year Plan period, our country made outstanding achievements in debris impact risk assessment, debris protection design and manned spacecraft protection[4], but still lagged behind foreign advanced countries in many aspects. In terms of satellite protection, the international development trend and the suggestions on domestic development are described below for different key techniques and research contents.

1. **Technique of ground-based hypervelocity impact test**

At present, the most commonly used ground-based UHS launcher is two-stage light gas gun. To simulate the impact effect of space debris more objectively and effectively, the development of new ground-based UHS launcher has always been the focus in the field of space debris protection. So far, the technology of 7 km/s impact, measurement and diagnosis of millimeter-sized spherical projectiles has become mature in China, but still lags behind the foreign advanced level. To this end, China needs to vigorously develop other new launching means and test techniques, for example, metal-foil electric explosion drive, magnetic drive, and three-stage gun technology based on two-stage gun, to improve the ground launching technology to the full coverage of 0–10 km/s.

2. **Numerical simulation technique and database construction for hypervelocity impacts**

In foreign countries, the method of numerical simulation is gradually used for analyzing the damage characteristics and failure modes of typical components or materials on satellites exposed to hypervelocity impacts. In addition, the numerical simulation process is constantly optimized and demonstrated to verify its effectiveness. The numerical simulation technique depends on the data of hypervelocity

---

[5] http://www.esa.int/Our_Activities/Space_Safety/Space_Debris/Hypervelocity_impacts_and_pro-tecting_spacecraft.

impact tests, which, however, is scattered and lacks the validity check. Therefore, the IADC pays attention to and promotes the construction of a hypervelocity impact characteristics database. The database construction undertaken by the EMI in Germany has made significant progress, containing a variety of hypervelocity impact test data on spacecraft protection structures and components that has been obtained in Europe in recent decades. China should strengthen the construction and unified application of a hypervelocity impact characteristics database for localized protective structures and materials and make a thorough research on the numerical simulation technology for UHS impact tests.

2. **Advanced protective structures and materials**

As the focus of spacecraft protection, the research on advanced protective structures and materials has never been interrupted either at home or abroad. The existing international cutting-edge high-performance protective materials include polyurethane foam, foamed aluminum, and polyolefin compounds. The performance of new foreign protective materials and structures has been not only enhanced gradually, but also developed towards multiple function directions, such as deployment ability, structuring, and space radiation resistance. In China, the spaceflight experiences should be based on to take the development and engineering application of polyurethane foam material and structure as the short-term development goal of spacecraft protection, and the development and engineering application of aluminum foam material and structure as the medium and long-term development goal of spacecraft protection. To support China's future spacecraft development, the R&D ability of domestic advanced composites research bases can be relied on to tackle key problems in the research of advanced protective materials and structures and meet the engineering needs.

4. **In-depth research on the damage and failure mechanisms of satellite components**

The research on the damage and failure mechanisms of typical satellite components needs a large number of ground-based tests and numerical simulations related to hypervelocity impacts, and the relevant conclusions can be used to guide the design of satellite protection. Therefore, to achieve the long life, high value and high reliability of spacecrafts, the in-depth research on the damage and failure mechanisms of satellite components exposed to hypervelocity debris impact should be carried out as a key task in the field of space debris protection.

5. **Debris cloud research**

From the perspective of spacecraft protection, the distribution characteristics of debris cloud in a given environment determine the penetration performance of debris cloud, thus determining the protective effect and spacecraft safety. Therefore, identifying the formation mechanism and penetration performance of debris cloud is helpful to guide the engineering design of spacecraft protection structure. In the past 30 years, the debris clouds were studied mainly in four aspects [5]: formation process of the debris clouds, distribution characteristics of the debris clouds, debris cloud models, and penetration performance of the debris clouds. The development trend of debris cloud research [5] is to build a debris cloud on the basis of single-debris feature statistics, instead of directly observing the macro characteristics of debris clouds. This change of cognitive perspective will lead the debris cloud research in the future. The measurement of debris cloud characteristics in numerical simulation will be based on debris instead of the previous SPH particles, so as to further analyze the change mechanism of debris cloud penetration limit, improve the whole research framework and guide the design and material selection of protective structures. The specific research directions can be summarized as follows [5]: (1) to analyze the formation process of debris cloud, especially the effects of shock wave propagation/attenuation and rarefaction wave in the projectile on debris formation; (2) to study the techniques of debris identification and feature extraction according to the distribution characteristics of debris cloud; (3) to improve the framework of debris cloud model and pay high attention to establishing the relationship between impact conditions and the overall distribution of debris cloud; (4) to quantify the penetration performance of debris cloud; (5) to optimize the design of sheet materials and more deeply analyze the protective effect brought by new materials; and (6) to further study the effect of phase change on the formation and distribution of debris cloud in a higher-speed impact, as well as the corresponding penetration performance of debris cloud.

In the field of space debris protection, foreign countries are not only doing the above work, but also actively developing the on-orbit impact perception technology [6]. Currently, the NASA and the JAXA are engaged in relevant research, hoping to realize the on-orbit detection and protection application combined with spacecraft engineering in the future.

6. **Protection needs transitioning from manned spacecrafts to satellites**

The protection of a manned spacecraft is very different from that of a satellite. On the one hand, a manned spacecraft (which can be understood as a large container) is considered failed when its cabin wall is perforated by space debris. The protection design for a manned spacecraft is based on the Possibility of No Perforation (PNP), that is, to assess the perforation risk of different parts of the spacecraft according to the size, impact velocity and impact angle of space debris and compare it with the given PNP. If the PNP of one area is lower than the given value, a protective structure should be added to the area to obtain a PNP greater than or equal to the given value. However, the case of a satellite is different. The performance degradation and failure thresholds of different components/subsystems of a satellite vary greatly under the impact of space debris. For example, solar wings are often perforated by space debris, which generally has no substantial impact on their function of power supply. However, the cut-off of main satellite cable by a debris impact will lead to the loss of satellite functions and even the failure of the whole satellite. On the other hand, the manned spacecraft has a regular geometric shape, so its main areas can be occluded. In comparison, the geometric structure of a satellite is much more complex, so many areas can not be occluded. Therefore, the design method based on PNP is not suitable for satellite protection. A top-down vulnerability study should be carried out for satellite to obtain the failure modes of satellite components/subsystems in a debris impact, so as to lay a foundation for the comprehensive assessment of satellite risks in the space debris environment, the identification of weak links and the targeted protection design. This is also the reason why the IADC is focusing its efforts on satellite vulnerability research. Of course, the viability of astronauts on the space station should also be evaluated.

## 6.7 Protective Engineering Application

### 6.7.1 International Space Station

The International Space Station (ISS) is the largest and most complex space project developed in the aerospace field in recent years. Its protective design requires a 10-year PNP of more than 81% or a perforation possibility of less than 19%. To this end, the ISS is equipped with more than 400 types of protective structures. Each module has different requirements for non-perforation reliability, surface area and operation time, and is provided with a different protection design according to its structure, impact probability and reliability requirement. The Micrometeoroid/Debris Protection System (MDPS) of the ESA's "Columbus" module was built by Astrium in France. It consists of 81 shields in two types, namely single shields and double shields. The single shield is a 1.6 mm-thick aluminum plate mainly used for protecting the cabin's sidewall and inner cone. The double shield is composed of a 2.5 mm-thick aluminum plate, 18 layers of resin-reinforced Kevlar fabric and 4 layers of Nextel fabric, with a centimeter-sized spacing between the aluminum plate and the stuffed layer. The double shield is mainly used for protecting the cabin's front face and outer cone. Both single and double shields are placed a few centimeters from the cabin wall to protect the pressurized cabin. The Japan Experiment Module (JEM) uses two types of shields: Whipple shield and stuffed shield. The main components of a shield include: aluminum mesh, NextelAF62, and Kevlar710. The distribution PNP of JEM is 98.37%, including 98.14% for pressurized module and 99.22% for experiment module. Through the above protection design, the JEM can meet the requirement of the design index. The wall of American Common Laboratory Module (CM) is composed of the 4.8 mm-thick A12219T87 plates with a total surface area of 133.8 $m^2$. Each column section is protected by a 2 mm A16061T6 plate with a spacing of 10.7 cm from the module wall. The protective spacing between frontal cone sections in the flight direction has been increased to 22.2 cm. The high debris-risk area on the side of the column section is protected by a more advanced stuffed Whipple shield. After the adoption of this protection design, the 10-year PNP of the CM is as high as 99.30%. The wall of Russian service module is composed of 1.6 mm (or 2 mm) AMG6 aluminium alloy plates with a total surface area of 140 $m^2$. Each module section is protected by the 1 mm-thick AMG6 aluminium alloy plates with a spacing of 5 cm from the module wall. After the adoption of this protection design, the 10-year PNP of the Russian service module is 98.37%. It is worth noting that the original PNP values of both the Russian service module and the core module (required to have a PNP of 98.63%) were lower than the ISS design requirements so that additional protective plates had to be installed for several times in the later years to reduce the risk of module failure caused by debris impacts. The ISS propulsion system consists of a main propulsion system and an auxiliary propulsion system, which are mainly used to provide an attitude control moment, orbital velocity correction, orbit maintenance and debris avoidance maneuvers. To ensure the special roles of propulsion system, each tank is not only physically separated from others, but also specially protected. The whole protective structure is basically made of aluminum material. The shield is 4-8 in. from the rear wall and is easy to remove. This facilitates both the replacement of the shield itself and the maintenance of the shielded devices.

### *6.7.2 "Tiangong" Series*

To protect China's long-life manned spacecrafts "Tiangong 1" and "Tiangong 2" from space debris impacts, the General Department of the China Academy of Space Technology (CAST) has developed MODAOST, a debris protection software package of international advanced level, carried out a large number of hypervelocity impact simulations and tests, and proposed the use of Whipple shield for local protection. Considering the single point failure of the radiator pipeline of "Tiangong-1", the CAST made major changes to the design configuration of the spacecraft with the support of hypervelocity impact test data, so that the impact resistance of the radiator pipeline was significantly improved without additional weight gain. The safe orbiting of "Tiangong-1" in 5 years from September 2011 to March 2016 and the safe orbiting of "Tiangong-2" in 3 years from September 2016 to July 2019 fully demonstrated China's self-developed debris protection design system and protection structure design technology. The advanced protective structure stuffed with basalt and aramid fiber is used in China's space station, with a protection capacity more than 50% higher than Whipple shield of the same mass.

### *6.7.3 Mars Exploration*

In addition to the protection design of traditional Earth-orbiting spacecrafts, the special characteristics of the environments in interplanetary space and in the vicinity of Mars should also be considered in the protection mission of Mars probe. The interplanetary space environment is not protected by the Earth's magnetic field, facing a magnetic field even weaker than the Earth's magnetic field. In this environment, solar wind can directly arrive at the probe surface, but their particles cannot enter the probe due to too low energy (electron: < 1 eV, positive ion: < 1 keV). The probe will be no longer affected by the Earth's radiation belt, but will be continuously exposed to galactic cosmic rays, which will become the main source of charged particle radiation during this operation stage. A major solar proton event may occur once after about 8 months. The probe will expose itself to a continuous high-vacuum environment. Therefore, more attention should be paid to the following characteristics and the corresponding protective measures should be taken in the development of a Mars probe.

1. **Total dose effect of ionization**

Since the Earth's radiation belt is not present, the total accumulated radiation dose will be almost zero if the solar proton event does not occur. The breakout of solar proton events has a certain probability and randomness. In this period and even in the next several Mars-orbiting months to several years, the Mars probe may encounter a very large solar proton event similar to that in October 1989. The radiation dose caused by such an event on the Mars probe is more than one order of magnitude higher than that caused by the Earth's radiation belt. The radiation dose still has little impact on electronic components and materials, but is the top priority for astronaut radiation protection in the manned Mars mission. This dose may or may not occur during this period. If occurring, the dose will be one-off and have little to do with the flight time.

2. **Single event effect**

In the Earth-Mars transfer stage and Mars-orbiting stage, the high-energy particle environment that the Mars probe is faced with can cause the single event effect (SEE). Except for the Earth's radiation belt protons (present in the initial launch stage of a Mars probe and in the orbital environment of an ordinary Earth-orbiting spacecraft), this environment mainly consists of galactic cosmic rays and solar cosmic rays and is basically similar to the environment that a lunar probe is faced with during its journey to the moon. In general, the SEE occurrence frequency at this stage is not much different from that of a conventional Earth-orbiting spacecraft or a lunar probe, when the use of electronic components particularly sensitive to single particles is not considered for Mars probe. However, the distance between the Mars probe and the Earth will gradually increase to 300–400 million km, so the maximum time delay for radio waves to be transmitted from the probe to the Earth can reach 17–20 min.

Due to this feature, the function of electronic devices will be interrupted once the SEE occurs in the probe. If the telemetry is relied on to detect and identify a fault and then send a remote control command for recovery, more than 40 min will be needed. If the fault is found in a key device closely related to the operation, attitude or critical action of the probe, the probe may be out of control during this period. Therefore, the method of SEE protection based on ground remote control is effective for an Earth-orbiting spacecraft or a lunar probe, but may not be effective for a Mars probe. The key to the SEE protection of a Mars probe is to ensure the self-diagnosis and self-recovery of on-orbit faults.

### 3. Solar radiation energy

As the distance between Mars and the sun is about 1.5 AU, the distance between the Mars probe and the sun will gradually increase from 1 to 1.5 AU during the journey from the Earth to Mars, and the solar radiation energy received by the probe will be gradually attenuated. At about 1 AU from the Earth, the solar radiation energy is about 1366 W/m$^2$. When the probe is near Mars, the solar radiation energy will drop to 500–700 W/m$^2$, as shown in Table 6.1.

When the Mars probe uses a solar array as energy supply, the solar array power will be greatly attenuated in the journey to Mars due to the continuous attenuation of solar radiation energy in the flight process to the minimum after the arrival at Mars. The power output of a solar array near Mars is almost half or less of that of another array with the same area near the Earth, which puts forward important constraints on the design of a solar array as the primary power supply system. With the decrease of solar radiation energy in the journey to Mars, the output power of the solar array will gradually decrease, calling for a need for shunting the large amount of electric energy initially stored in the probe near the Earth. This poses a challenge to the design of power controller, and may require the measures taken for both the power controller and the probe as a whole. In addition, since the solar radiation energy drops to the minimum near Mars, the external heat flux obtained by the probe near Mars will be much lower than that near the Earth and may not be enough to meet the thermal control requirement of the probe. In this case, special measures (such as heat collector and gas insulation) different from those for an Earth-orbiting spacecraft or a lunar probe need to be taken.

### 4. Vacuum cold-welding and dry friction

Similar to an Earth-orbiting spacecraft, the Mars probe is immersed in a high-vacuum environment with a vacuum of $10^{-14}$ Pa during the whole Earth-Mars transfer process. Many components of the probe, such as parachute deployment device, moving parts of the buffer engine, lander, and moving parts of the rover, will take action or be released only after arriving at the region near Mars or landing on Mars. In the long-time Earth-Mars transfer stage, they are pressed and inactive and thus are

**Table 6.1** Solar radiation attenuation during Earth-Mars transfer in typical Mars missions

| Item | Distance to the sun/AU | Solar radiation energy/(W·m$^2$) |
|---|---|---|
| Average distance between the Earth and the sun | 1 | 1366 |
| Maximum value in a mission | 0.98 | 1422 |
| Average distance between Mars and the sun | 1.52 | 591 |
| Maximum value in Mars orbit (at the minimum perihelion) | 1.38 | 715 |
| Minimum value in Mars orbit (at the maximum aphelion) | 1.67 | 491 |

prone to vacuum cold-welding or dry friction, which will have a significant impact on their deployment, movement or release after arriving at Mars, and even lead to a mission failure. Therefore, vacuum cold-welding and dry friction are the important secondary effects of the Earth-Mars space environment under which the probe should be protected.

The characteristics of the space environment near Mars are as follows:

1. **Martian dust and dust storms**

The surface of Mars is covered with dust. In the thin and dry atmosphere, the dust will be kicked up into the atmosphere by winds, and can turn into regional or global dust storms when certain conditions are met. Such dust storms often occur in the tropics and southern hemisphere (in summer) of Mars. A regional dust storm typically lasts for a few weeks and can cover a large area. Global dust storms occur on Mars almost once every Martian year and can last for several months. A global dust storm usually starts from the southern hemisphere when the Martian perihelion, namely the southern hemisphere of Mars, is in summer.

Martian dust storms could affect the power systems of the lander and rover that land on Mars. The dust storms have the strongest weakening effect on shortwaves (for example, blue light) in solar radiation. Therefore, in the design of solar arrays of the lander or rover, the technological parameters of solar arrays should be adjusted to better respond to red light and infrared rays. A dust storm will prevent the entry of solar radiation and reduce the average and maximum temperatures of the Martian surface, while hindering the infrared radiation of the Martian surface and thus raising the minimum temperature of the Martian surface. These changes apply constraints on the thermal control design of Mars lander and rover.

Once falling on the lander or rover, the Martian dust floating in the wind can block the sunlight incident on the solar array, thus reducing the power generated by solar array. The dust can shade the camera or optical sensor lens, affecting the camera imaging or the sensor operation. The dust can also enter the moving parts (such as bearings), causing various adverse effects (e.g. motion stagnation). Therefore, the relevant components and devices on the probe need to be well dust-protected to ensure normal operation of the probe on the surface of Mars.

2. **Radiation of charged particles**

The radiation of charged particles on the surface of Mars is relatively simple. Since Mars has no global magnetic field, there are no radiation belts of charged particles on its surface. Therefore, if a large solar proton event is not encountered, the radiation dose received by the probe will be almost zero during the Mars-orbiting period and the period of Mars surface mission, regardless of the mission period length. Galactic cosmic rays, which continuously irradiate the Mars lander and rover, are the main source of SEE radiation. Since the distance between the Mars probe and the Earth is too far, the key to the SEE protection is still autonomous processing, rather than relying on the ground intervention to long-delay radio wave propagation. Similarly, the surface charge/discharge effect and internal charging effect of an Earth-orbiting spacecraft will not occur on the surface of Mars due to the absence of thermal plasma and high-energy electrons on Mars.

The Mars probe will go through the Earth's magnetosphere, interplanetary space, and different space environments around and on the surface of Mars. Depending on the mission characteristics and probe design characteristics, the effect of space environment and the protection priority will change at different operating stages. The conventional space environment adaptability of the Mars probe to total ionization dose, Earth's atmosphere, Earth's magnetic field, vacuum and other effects is basically similar to that of a conventional Earth-orbiting spacecraft. The special space environment protection for a Mars mission mainly includes the following aspects: (1) autonomous detection of and recovery from single particle effect in a long-distance flight; (2) influence of solar radiation energy reduction (when the probe is far away from the sun) on power supply design and thermal control design; (3) low-pressure discharge caused by low CO2 pressure on the surface of Mars, and other related problems; (4) influence of Martian atmosphere and wind on orbit design and landing safety in the Mars-orbiting and EDL processes; (5) influence of light shading and solar radiation occlusion (caused by Martian dust storms) on power supply design and thermal control design; and (6) influence of Mars landform uncertainty on landing safety and on the rover travel and obstacle avoidance.

## References

1. Gong, Z., Zhao, Q., Li, M., et al. (2019). Frontiers and prospects of debris protection research. *Space Debris Research, 19*(3), 2–13.
2. Qu, G., & Han, Z. (2005). General conception of spaceborne passive protection technology against space debris. *Spacecraft Engineering, 14*(2), 1–7.
3. Li, Y., Shen, H., Wang, W. (2011). Numerical simulation technology of space debris hypervelocity impact and its application. In: *System Simulation Committee of Chinese Society of Automation, and Simulation Technology Application Committee of China Simulation Federation. Proceedings of the 13th Annual Conference on Chinese System Simulation Technologies and Their Application* (Vol. 4). System Simulation Committee of Chinese Society of Automation, and Simulation Technology Application Committee of China Simulation Federation
4. Gong, Z., Han, Z., & Pang, B. (2010). Research status of space debris protection in China and development suggestions. *Spacecraft Environment Engineering, 27*(1), 24–31+5.
5. Han, Z., & Pang, B. (2012). Recent progress in space debris protection research. *Spacecraft Environment Engineering, 29*(4), 369–378.
6. Di, D., Chen, X., Wen, K., & Zhang, C. (2018). Progress in the research of debris cloud generated by hypervelocity collision. *Acta Armamentarii, 39*(10), 2016–2047.

# Chapter 7
# Design of Space Debris Mitigation and Removal

## 7.1 Overview of Space Debris Mitigation

The harm caused by space debris to the environment is a concern of the whole world. On the one hand, it is possible for every spacecraft to form the space debris and for almost every country engaged in space activities to produce the space debris. On the other hand, the space debris poses a threat to all the orbiting spacecrafts. Therefore, how to effectively control the growth of the amount of space debris and minimize the generation of space debris is the premise for the peaceful use and protection of outer space resources. This is not only the joint goal of all space-faring countries, but also the international obligation and mission that each space-faring country must undertake.

The ultimate goal of space debris mitigation is to restrain the deterioration of space environment, protect the space environment resources, and realize the sustainable utilization of space environment resources. In general, the debris mitigation measures fall into two categories:

(1) Preventive mitigation measures; and
(2) Removal mitigation measures.

The former aims to minimize the generation of space debris in the period from the beginning of a mission to the end of the spacecraft's life, while the latter aims to remove the space debris already in orbit.

### *7.1.1 Technical Guidelines and Standards for Mitigation*

To reduce the generation of space debris, the countries or organizations engaged in space activities have issued their own debris mitigation guidelines or standards to curb the growth of space debris. The existing international mitigation guidelines and



J. Zhang et al., *Space Debris Research*, https://doi.org/10.1007/978-981-96-9570-6_7

standards mainly include the debris mitigation standards issued by NASA, IADC, ESA, NASDA, RSA and CNES.

1. **Debris Mitigation Standards Issued by NASA**

The first debris mitigation standard issued by NASA was the Safety Standard 1740.14, which was developed based on the NASA's management directive 1700.8 on "Policies to Limit Space Debris Generation". After the safety standard was formally adopted, the NASA required all programs and projects to provide the debris assessment reports during the design and development phases. The report should include five main aspects:

(1) On-orbit collision evaluation

As the large space debris can cause a destructive damage to spacecraft, a piece of large-sized debris colliding with an orbiting spacecraft will produce more small-sized fragments, which will pose a threat to other orbiting spacecrafts. Therefore, the NASA's safety standard requires the evaluation of the probability of a collision between the spacecraft and large debris during the mission, and also requires an analysis of the potential damage caused by various types of potentially dangerous small debris, as well as the consideration of various types of collisions that might be foreseen at the beginning of the mission.

(2) Principle of post-mission treatment

According to the safety standard, the final stage of a launch vehicle or a satellite that is about to fail needs to be moved to a predetermined treatment orbit or an orbit with a life span of less than 25 years to minimize the time of staying in an effective orbit and reduce the threat to other orbiting spacecrafts. In this process, the feasibility of the treatment scheme should be evaluated and analyzed to ensure the effective and normal implementation of the treatment scheme.

(3) Size of on-orbit disintegration fragments

The on-orbit disintegration is either an accidental disintegration or a man-made disintegration. Accidental disintegration is generally the spacecraft disintegration caused by an accidental impact or by the accidental explosion of the spacecraft itself. Man-made disintegration is the purposeful disintegration realized under artificial control. The safety standard requires to control the amount of the debris larger than 1 mm generated during any disintegration event as well as the probability of accidental spacecraft disintegration.

(4) Generation and release of debris during mission

The carrier rockets and spacecrafts are bound to produce debris during the launch or normal operation. Moreover, they have some components that must be abandoned. Therefore, the safety standard clearly defines the mission debris as large components (such as fairing) in addition to the uncontrollable debris generated during the spacecraft launching and orbiting, while limiting the amount of the released debris larger than 1 mm and its on-orbit life.

(5) Debris reentry limitation

The safety standard sets strict limits on the size and amount of uncontrollable reentry debris produced by a spacecraft.

2. **IADC's Space Debris Mitigation Guidelines**

The *Space Debris Mitigation Guidelines* are the space debris mitigation standards promulgated by the IADC Mitigation Technology Group (Technical Working Group IV) in 2002. The guidelines are based on the analysis and assessment of the space debris mitigation standards and relevant technical measures formulated and promulgated by the major space-faring nations worldwide. The guidelines cover the whole process from the initial phase of project approval to the on-orbit operation and post-mission treatment of spacecrafts and vehicles. To further improve the guidance and operability of the Guidelines, the IADC Mitigation Technology Group later issued the *Technical Support for Space Debris Mitigation Guidelines* as a supporting document to the Guidelines.

The IADC's *Space Debris Mitigation Guidelines* consist of four parts: technical standards for mitigation, technical measures for mitigation, mitigation management labeling, and mitigation effectiveness assessment. The contents of the Guidelines mainly include the following four aspects:

(1) Preventing the collision of spacecrafts in orbit;
(2) Reducing the debris amount generated during a normal launch;
(3) Reducing the risk of spacecraft explosion or disintegration in orbit; and
(4) Formulating the principles for the disposal of spacecrafts after mission termination.

3. **ESA's Debris Mitigation Standard**

The ESA developed a safety standard for the European countries or institutions engaged in space activities. The standard mainly includes the requirements for the design, management and operation of space activities. The standard also requires that all projects and programs provide viable debris mitigation plan and measures and assign a person to oversee and implement the mitigation plan throughout the process. The standard mainly includes the following four aspects:

(1) Post-mission treatment

The standard requires that a spacecraft be passivated at the end of its mission and remain passivated during deorbit and orbital transfer. For example, all energy of the batteries in the spacecraft should be released and their power circuits should be disconnected at the end of the mission. The residual pressure in the booster tank should be kept below 50% of its critical pressure to ensure that the tank will leak only rather than explode if punctured or damaged.

(2) On-orbit disintegration

The ESA's standard clearly prohibits the man-made on-orbit disintegration of spacecrafts and requires no active debris generation from the onboard materials during service. The standard also requires the explosive probability of spacecraft components (such as propulsion devices, booster components and batteries) to be less than 0.01%, while putting forward requirements for the use of solid propellant.

(3) Limitation of mission debris

According to the standard, the final stage of a carrier rocket should not produce long-life space debris, and all types of components of a spacecraft (for example, explosive bolts, springs, cable accessories, and covers of detection instruments) should be tethered to the spacecraft to avoid their free movement in space. If the requirement of debris limitation cannot be met, the amount, size, orbital parameters and lifetime of debris should be preliminarily estimated.

(4) Debris reentry

According to the standard, the reentry debris should not produce chemical or radioactive substances harmful to the Earth's environment, and the risk of its damage to the ground should be less than 0.01%.

4. **NASDA Standard**

Japan's space debris mitigation standard, NASDA-STD-18, was developed by Japan's National Space Development Agency (NASDA) in 1996. The standard also sets requirements and restrictions on mission debris limitation, on-orbit disintegration, post-mission treatment and debris re-entry, which mainly involve four specific aspects:

(1) Reducing the amount of debris generated and released by the orbiting spacecrafts;
(2) Shortening the residence time and life of the final stage of a spacecraft or rocket in orbit after the mission completion;
(3) Reducing the probability of on-orbit disintegration; and
(4) Specifying the orbital region to which a GEO spacecraft can be transferred after the mission completion.

The NASDA's mitigation standard provides some data requirements that can be used for reference. For example, a spacecraft with an orbital altitude higher than 1500 km is required to enter a graveyard orbit at an altitude of 1700 km after the completion of its mission. A spacecraft with an altitude of 1000–1500 km should have an orbital life shortened as much as possible or should be sent into the predetermined graveyard orbit. A spacecraft with an orbital altitude of lower than 1000 km should be descended with the help of propellant consumption or atmospheric drag and should have an orbital life limited to less than 25 years.

5. **RSA Standard**

The RSA standard is a space debris mitigation standard formulated by the Russian Federal Space Agency (RSA) in 2000. Its main contents are as follows:

(1) Artificial self-destruction of spacecraft is not allowed;
(2) The remaining propellant in a carrier rocket should be drained off after the mission completion to reduce the pressure of fuel tank;
(3) At the end of the mission, the spacecraft should exhaust its remaining propellant and prevent the separation of its components;
(4) The batteries on a spacecraft at the end of life (EOL) should be passivated;
(5) The propulsion unit of an apogee engine should not be jettisoned.

6. **CNES Standard**

The CNES's debris mitigation standard was developed in 1999 by the Centre national de la recherche scientifique (CNES) in France. This standard mainly sets requirements and restrictions on five aspects: on-orbit disintegration, on-orbit collision, post-mission treatment, mission debris limitation, and debris reentry. Its contents are basically consistent with NASA's safety standard. According to some of the specific mitigation targets set in the standard, the reliability of post-mission treatment solution of a spacecraft is required to be more than 99%; as few components as possible other than the payload are allowed to enter the orbit, for example, only the final stage of a launch vehicle is allowed into orbit.

The above-mentioned space debris mitigation guidelines and standards formulated by different countries or organizations have basically the same contents. On the whole, they consider the space debris mitigation from four dimensions: limitation of mission-related debris, on-orbit disintegration, debris re-entry, and post-mission treatment. However, they are slightly different from each other in some aspects.

In terms of the limitation of mission-related debris, only the NASA requires assessment and analysis of the amount, size and orbital lifetime of the debris larger than 1 mm. However, the NASDA considers it difficult and impractical to evaluate and analyze small debris, thus setting only the limit on the amount of debris. The CNES's mitigation standard sets restrictions on the number of objects into orbit except the payload and on the type of propellant and materials used by the spacecraft, in order to minimize the generation of debris from all aspects. In the RSA standard, there is no specific limit on the amount of mission-related debris.

Because on-orbit disintegration is a major cause for space debris generation, all mitigation standards other than NASA standard explicitly prohibit the man-made self-destruction of spacecrafts. The NASA standard does not prohibit man-made on-orbit disintegration, but puts forward clear requirements for the amount of disintegration debris and requires the probability of on-orbit disintegration to be lower than 0.01%. The NASDA and RSA standards have basically the same mitigation requirements, requiring to minimize the possibility of disintegration and investigate various possible causes for disintegration, such as propellant, battery and other potential factors. The CNES also requires the probability of accidental disintegration of a spacecraft in orbit to be less than 0.01%, while requiring the storage tank to be passivated in the process of deorbiting or orbital transfer to prevent its explosion.

### 7.1.2 Preventive Mitigation Measures

1. **Prevention of On-Orbit Collisions**

Prevention of on-orbit collisions is a preventive technical measure designed to avoid collision with large space objects through monitoring, warning and orbital maneuvering, thereby reducing the generation of new debris.

2. **Reuse**

If the EOL spacecraft and the final stage of a carrier rocket become space junk due to the depletion of their propellant, their economic benefits will be reduced. Although their fragmentation in space can be prevented by sending them into the atmosphere for incineration, such treatment may cause pollution to the atmospheric environment from the perspective of long-term environmental protection. Therefore, the scholars at home and abroad have paid attention to the reuse of carrier rockets and spacecrafts in recent years. For example, the existing single-use launch vehicles may be replaced by the single-stage rockets that can be launched repeatedly or by space transportation systems such as new space shuttles. Technologies such as on-orbit maintenance, refueling, and recovery of space platforms can also be used to extend the service life of on-orbit satellites so as to reduce hardware consumption and the generation of space debris.

3. **Transfer to Graveyard Orbit**

Graveyard orbit, also called junk orbit, is an orbit with less practical value for space activities. The transfer of satellites to a graveyard orbit is an effective preventive technology of on-orbit debris population mitigation. According to the different orbital altitudes, the graveyard orbits can be divided into three types: low-orbit graveyard orbit, high-orbit graveyard orbit and GEO. For a satellite on GEO, the graveyard orbit is required to be at least 300 km higher than that of GEO to avoid the collision between the satellite and other satellites operating normally on GEO or GTO when its mission is completed. When the propellant of the GEO satellite is exhausted, the satellite will be unable to keep stationary relative to the pinpoint and will drift under the perturbation of solar and lunar gravity and Earth oblateness. The orbital inclination of the satellite will change periodically with a 53-year cycle. In this process, the maximum inclination can reach 15°. The satellite will pass through the GEO plane twice a year with a relative velocity of about 800 m/s. This poses a great threat to other satellites operating normally in GEO orbit.

Sending a GEO satellite into the graveyard orbit is a relatively feasible method to prevent it from becoming a safety threat after the completion of its mission. In addition, most of the countries engaged in space activities have reached a certain consensus on the altitude of the graveyard orbit and the strategy of orbital transfer, that is: a circular orbit 300 km higher than GEO is selected as the graveyard orbit, and a velocity increment of about 3.65 m/s is provided for each orbital transfer to raise the orbital altitude by 100 km through multiple transfers. The propellant consumption in this process is equivalent to the propellant amount needed to keep the satellite stationary relative to the pinpoint for one month. This method has been applied nearly 100 times to the satellites operating in the GEO region, and has proved very effective. However, the orbital transfer needs the consumption of a certain amount of propellant, which will shorten the satellite's service life. To solve this problem, the whole international community is devoted to the study of fuel-saving propulsion methods to achieve a reliable transfer of satellites to the junk orbit.

4. **Tether**

Tether is a technical means to retain the space debris or discards generated during the launch and operation of a spacecraft by designing the relevant mechanisms. On the one hand, the generation of debris or discards needs to be minimized during the spacecraft mission. On the other hand, the parts that must be jettisoned, such as rocket fairing, lens protection cover, clamp band of spacecraft-rocket separation device, and instrument probe, are required to be connected to the spacecraft or the rocket's final stage through a tether system after being released, so as not to move on the orbit at will.

5. **Passivation**

Passivation is a preventive measure to treat the objects that may cause the accidental explosion of a spacecraft, which include batteries, propellants and high-pressure gases in the spacecraft or the final stage of carrier rocket. By pre-treating such objects, all factors that may cause explosion in orbit can be rooted out. Therefore, passivation is also called "energy dissipation".

At least 184 cases of on-orbit object disintegration have been recorded, each of which produced a large amount of space debris. As far as the debris that can be detected by ground observation system is concerned, the debris amount generated by the disintegration events accounts for 37.7% of the total debris amount. On June 3, 1996, the launch vehicle "Pegasus" exploded, producing 709 pieces of debris, the largest amount of explosion debris ever recorded.

According to the different objectives, the passivation methods are generally divided into three types: passivation of moving parts, battery passivation and propellant passivation.

The passivation of moving parts is specially applied to the movable parts of the spacecraft, such as control torque gyro. According to the passivation requirement, all moving parts of the spacecraft should stop moving after the completion of the mission, in order to prevent their accidental failures that will lead to the generation of debris. Since all the onboard moving parts need to be powered before continuing to work, the simplest passivation solution for them is to design a device in their circuit to cut off their power. When the spacecraft has finished its mission, the moving parts should be powered off by a command and then stop moving.

The battery passivation requires the spacecraft to release the energy stored in batteries and disconnect the power supply circuits after the completion of its mission. When the energy storage components like batteries are knocked or damaged in the working process, they are extremely prone to explosion and then produce space debris. Therefore, all the energy stored in batteries must be released and their power supply lines must be disconnected after the completion of the spacecraft mission. The method of battery passivation is also to design the devices in the control circuit to discharge and disconnect the batteries. After the mission completion, a preset instruction is sent to stop the power supply from the batteries.

According to the requirement of propellant passivation, the propellant or high-pressure gas left in the tank after the separation of the final-stage rocket from the spacecraft must be released in time to eliminate the potential risk of explosion. The propellant can be removed in two ways. One is the direct discharge of propellant. The tanks with surface tension are most often installed in the existing spacecrafts to facilitate the discharge of the remaining propellant or high-pressure gas. The other way is to re-ignite the engine and burn the propellant out. The former method is usually a balanced discharge method with little influence on the movement of the rocket body. In contrast, the latter method will produce a larger thrust and affect the orbit of the rocket. In the latter method, the collision with other spacecrafts operating normally in orbit should be avoided.

### *7.1.3 Removal Mitigation Measures*

1. **Deorbit**

Deorbit is a technical measure to transfer the spacecraft and the final stage of carrier rocket to another orbit at the end of their mission in order to shorten their residence time in the original orbit. Depending on whether the deorbit is powered or not, the deorbit methods can be classified into two categories: powered deorbit and unpowered deorbit.

Powered deorbit is a process that the launch vehicle or spacecraft relies on its own power to leave the original orbit. As far as the launch vehicle is concerned, some propellant is left in the vehicle after sending the spacecraft into orbit. By using the remaining propellant for re-ignition, the vehicle can leave its original orbit as long as the thrust direction can be correctly controlled. More residual propellant may be required for the vehicle to return to the Earth and less for the vehicle to enter the LEO or a junk orbit. The upper stage of the "CZ-2C" rocket also successfully achieved an active deorbit after launching the American "Iridium" satellite.

Unpowered deorbit is mainly to make space debris leave orbit and fall quickly by using space environment resources. To this end, two kinds of space environment resources are available: atmospheric resistance and the drag force generated by geomagnetic field [1].

2. **Recovery**

Recovery is a technical measure to capture space debris and bring it back to the ground by using a spacecraft capable of round-trip flight. Currently, only the U.S.

space shuttle and Russia's "Soyuz" manned spacecraft are capable of flying to and from the low Earth orbit (LEO). However, due to too high launch cost and too low safety performance, the space shuttle only recovered a limited number of space fragments. Some recovery missions, such as the recovery of the "long-term orbit exposure device" and the solar panels of "Hubble" Telescope, were designed for on-orbit collision tests and spacecraft maintenance, rather than space debris removal. Meanwhile, because the orbital altitude that the space shuttle and the manned spacecraft can reach is limited to generally less than 600 km, the recovery technology is only suitable for removing the LEO debris in theory.

1. **Collection**

Collection is a removal measure that uses an orbital collector to collect and remove tiny space debris. For example, a spherical collector with bubble-like holes on surface and with a diameter of thousands of meters is placed in orbit. When small fragments come into contact with the collector, they may be sucked into the holes or be slowed down by collision and then descend. But an ordinary collector is not capable of maneuvering and orbital transfer, and thus can hardly avoid a spacecraft normally operating. In addition, even if the collector is equipped with control system and propulsion system, a reliable identification technology is still needed to distinguish an operating spacecraft from space debris to prevent the occurrence of wrong collection. This puts forward very high requirements for the technical performance of the collector, resulting in high cost. Therefore, the feasibility of collector needs to be further studied.

2. **Burning**

Burning is a removal measure that uses a high-power laser equipment to irradiate the space debris, which is then powdered, melted or even vaporized, leading to rapid orbital attenuation. The burning effect is affected by not only the working mode and power of the laser, but also the size and material of the space debris, as well as the altitude and inclination of the orbit where the debris is located. Therefore, this technology is only applicable to the small-sized debris in LEO.

## 7.2 Necessity of Active Removal

As the possibility of collision between space objects increases with the frequency of space activities, the amount of space debris is constantly increasing. The existing reports show that the amount of space debris is increasing at a rate of 10% per year [2], which is expected to last for at least 50 years. If the debris is not actively removed, the future space activities will inevitably be affected. For the space debris in LEO, its orbital descending relying only on atmospheric drag may last for more than 20 years. If the orbital altitude exceeds 200 km, the atmospheric drag will drop sharply, resulting in a longer descending time. Figure 7.1 simulates the debris

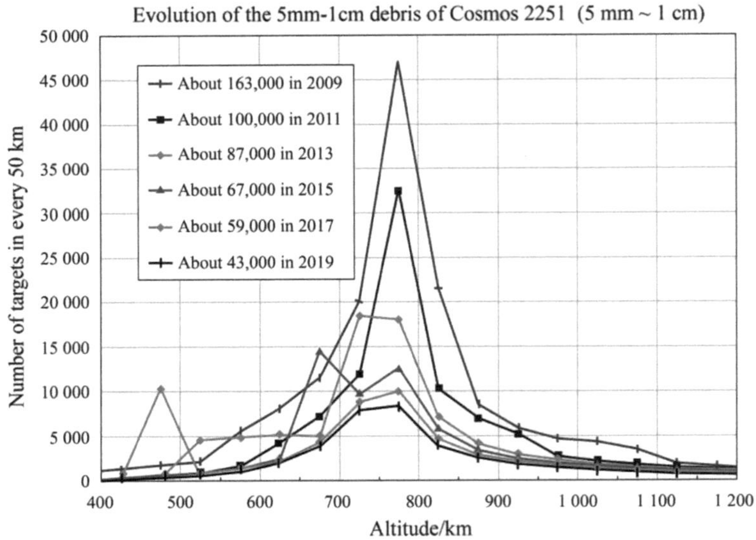


**Fig. 7.1** Altitude distribution simulation for the 5 mm–1 cm debris generated by "Cosmos 2251" collision (see the colored illustration)

altitude distribution generated by the collision between Iridium 33 and Cosmos 2251 in 2009. The simulation results show that under the combined action of atmospheric drag, solar and lunar gravity, solar radiation pressure and other perturbation forces, the space debris with the size of 5 mm–1 cm will rapidly attenuate and descend to a low altitude within a given altitude range of less than 1200 km.

The simulation results show that the automatic orbital descent of space debris can be realized by relying on environmental perturbation, which, however, is only effective for the space debris on LEO. In addition, this passive descending usually takes a long time. For the space debris on HEO, the time of orbital descent will be greatly extended due to the reduction of atmospheric drag, so the HEO debris basically cannot rely on this method for orbital descending. Therefore, it is necessary to explore the active removal technology to clean up space debris, so as to reduce the amount of debris, avoid the generation of more debris and thus Kessler effect due to collisions, and lower the risk in future space activities.

In general, the technology of active debris removal has a high application value in ensuring the safety of future space activities and governing the space environment. This is mainly reflected in the following five aspects:

(1) Applying the removal technology to on-orbit services. Since most of the objects to be removed are non-cooperative targets with strong uncertainty, quite a lot of problems need to be considered in the process of removal. However, if the objects are cooperative targets, this technology can be well applied to the on-orbit service activities theoretically. For example, the laser displacement technology can supplement energy to the target spacecraft through laser energy

transmission. The technology of capture can also be used in orbital transfer, dragging the spacecraft that failed to enter orbit into the predetermined orbit to achieve space rescue or extend its service life.

(2) Ensuring the safety of space activities. This is also the fundamental motivation for the exploration of active removal technology. According to reports, the International Space Station (ISS) has been hit by space debris and gas has leaked from the station. Although the ISS structure is designed with a capability of withstanding a collision with small debris, the station still mainly relies on orbital transfer to reduce the probability of colliding with large dangerous debris. Moreover, each orbital transfer requires a high cost, and cannot be realized in a short time. The active removal of these threatening fragments can not only greatly improve the viability of the station, but also greatly reduce its operating costs [3, 4].
(3) Promoting the development of more new space technologies. Because the technology of on-orbit debris removal is very complicated, the large-scale on-orbit debris removal based on the existing technology still needs a long way to go. However, some of the more advanced technologies have been proposed during this process, for example, platform maneuvering and on-orbit operation, space environment monitoring and application, high-energy and high-precision laser emission, as well as space detection and identification [5].
(4) High military value. Since most of the targets to be removed are non-cooperative targets without preset identifier of friend or foe (IFF), the removal technology can theoretically be used to remove enemy targets from the military point of view. This is one of the important reasons why the United States attaches great importance to the development of the removal technology.
(5) Huge commercial benefits. Since the orbit resources are limited in space, the occupation of an orbit of high practical value by space debris is likely to cause huge economic losses. Therefore, the removal of space debris may be of high commercial value. According to the *Reference News* released on November 13, 2000, the president Vitaliy Lopota of RKK Energia said that the use of spacecraft for space debris removal had a high potential commercial value because the GSO resources were limited and the fixed points for placing satellites were being gradually used up.

To sum up, the space debris has posed a serious threat to current and future space activities, thus becoming an urgent problem that needs to be solved. The space activities play an important role in human development, but how to effectively utilize and protect the space environment while carrying out space activities is a technical problem. If allowed to grow without being controlled by any measures, the space debris will likely become a huge obstacle to human spaceflight in a few decades. Therefore, it is of great theoretical value and practical significance to study and develop the methods and technologies of space debris removal.

## 7.3 Goals and Strategies of Active Removal

Due to the different functions and types of the launched spacecrafts, the amounts, types, sizes and material characteristics of space debris generated by these spacecrafts are also different. The first step of space debris removal is to determine the object to be removed. If the targets, requirements and purposes of the removal missions are different, the removal sequences will also be different.

Since the orbits of a practical value are limited, the space debris is distributed in a concentrated manner, mostly in some special orbits. As can be seen from the literature, more than half of the space objects orbiting the Earth are near the LEO. According to environmental studies, the instability of space debris groups will also be mainly reflected in LEO in the coming centuries [6]. In higher orbits, although atmospheric drag is less helpful in removing debris, the debris tends to accumulate more slowly because space activities are less frequent than in lower orbits. However, due to the special characteristics and resource scarcity of geosynchronous orbit and solar synchronous orbit, these orbits often have great commercial and scientific values and thus a very high value in debris removal [7]. Therefore, the removal of the debris on such orbits should be more urgent.

The space debris is commonly classified by size [8], as shown in Table 7.1.

The threat of space debris to a spacecraft often depends on the debris size. Generally speaking, at the same orbital altitude, the larger debris will have greater kinetic energy due to its larger mass, thus producing greater destructive power in the event of collision. In addition, it may also contain dangerous parts such as batteries and fuel, which constitute one of the potential sources of smaller debris [10]. Although the amount of such debris is small and the probability of colliding with it is relatively low, such a collision is often fatal. Therefore, the collision with such debris should be avoided first before considering its removal. As such large hazardous debris has a great momentum, its destructive power will overwhelm the current anti-collision means and can only be effectively coped with by avoidance.

**Table 7.1** Classification and characteristics of space debris [9]

| Category | Type | | |
|---|---|---|---|
| | Small debris | Middle-sized (dangerous) debris | Large debris |
| Size/cm | < 1 | 1–10 | > 10 |
| Hazard | Little or no damage | Local or global damage | Complete destruction |
| Motion information tracking | Not tracked, but statistically analyzed | Some of them are tracked | All are tracked |
| Quantity estimation/ PCS | > 100 million | About 500,000 | > 21,000 |
| Mass ratio relative to the total mass /% | — | 5 | 95 |

The large-mass debris has strong destructive power, but after all is in the minority. Most of the debris in space is small. Although the momentum of small debris is limited, such debris has a huge amount and brings a great difficulty in tracking it and very low actual benefits in removing it. Meanwhile, most of the small debris is concentrated in LEO. Therefore, its removal is not considered at the current stage. To deal with such debris, the spacecraft generally takes self-protection measures, such as anti-collision reinforcement, to reduce its impact [11].

In order to remove space debris, a variety of debris removal schemes have been put forward currently. By classifying space debris according to different characteristics, the removal measures suitable for different types of space debris have been proposed. As can be seen from Fig. 7.2, the removal methods are mainly divided into four types: laser-aided removal, push-off deorbiting, drag-increase deorbiting, and capture deorbiting, which deal with different debris sizes. On the whole, the laser-aided removal [12] is usually only applicable to the debris smaller than 10 cm due to the limitation of space environment and laser emitter power. The push-off deorbiting driven by ion beam or artificial atmosphere is usually applicable to the 10 mm–20 cm debris and is difficult to deal with too small or too large fragments due to the limitation of this deorbiting mechanism. The push-off deorbiting driven by solar sail requires the installation of solar sail on the target, so the debris has to be large enough and needs to operate on a high orbit. The four methods of drag-increase deorbiting (foam method, fiber method, dust method and expansion method) only deal with the LEO objects due to the mechanism limitations. Except in the dust method, the size of the objects in the other three drag-increasing methods should not be too small, otherwise the form, fiber or expansion substances can not attach to the objects to achieve the purpose of drag-increase deorbiting [12–14]. In the capture deorbiting, debris is captured and removed from orbit by means of rope net or mechanical arm, so the size of debris should not be too small [15].

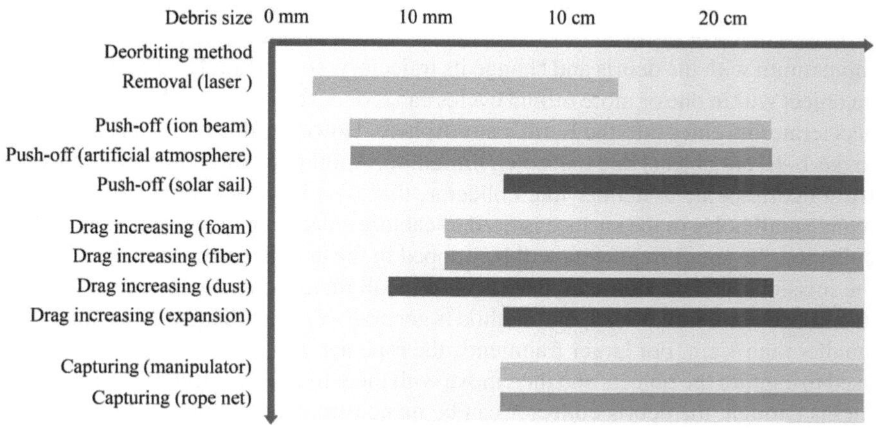


**Fig. 7.2** Application scopes of different removal methods (see the colored illustration)

## 7.4 Design of Technical Schemes

### 7.4.1 Target Screening

According to statistics, there are now about 16,000 cataloged pieces of debris and about 500,000 pieces of 1–10 cm debris in outer space, which are difficult to be tracked for the time being. The average amount of debris per cubic kilometer is small in the orbital space near the Earth, where only a limited number of orbits are of practical value. However, most of the debris is concentrated in GEO and LEO, which constitute one of the orbital ranges with the most frequent space activities and the highest practical value. Thus such debris poses a great threat to the current spacecraft operation.

### 7.4.2 LEO Debris Removal

LEO is a range of Earth orbits at an altitude of less than 2000 km. Since most of the spacecrafts are running in LEO, most of the debris is also generated in LEO. The new debris generated in LEO poses a new potential threat to the spacecrafts in LEO. If the debris is not removed, the space environment in LEO will enter a vicious circle. The LEO debris is desired to automatically descend under atmospheric drag. With the increase of drag action time, the kinetic energy and altitude of the debris decrease gradually. Finally, the debris falls from orbit into the atmosphere and burns up. The time of orbital descending is a function of the altitude of space object. The higher the orbital altitude, the longer the descending time. In most of the LEO debris removal technologies, the debris orbit altitudes are reduced by increasing the atmospheric drag. In addition to atmospheric drag, the laser beam technique is also proposed to lower the orbit of smaller debris. This technique uses lasers to exchange momentum with the debris and change its trajectory, so repeatedly firing the lasers to an object within one or more orbital cycles can reduce the object's orbital altitude and accelerates its entry into the Earth's atmosphere. However, this method is applicable to the 1–10 cm objects due to its own limitation. Another method for removing small LEO debris is the container-like collector, that is, a large spherical collector with many small holes in the surface is used to capture small fragments. When hitting the collector, the small fragments will be trapped in the hole net structure. At the end of the mission, the collector can leave orbit with all the captured fragments. However, due to structural limitations, this method is generally only applicable to the fragments smaller than 1 cm. For larger fragments, the rope net, robotic arm or tentacle can be used to capture the objects and then move with them to a lower orbit. If the remaining fuel is enough, the debris collector can be maneuvered again to collect other debris targets.

### 7.4.3 GEO Debris Removal

GEO is another range of Earth orbits with high practical value, and is located in a narrow region near the equator at an altitude of 35,786 km above the Earth. The GEO has no Earth atmosphere, thus providing no possibility of relatively efficient natural attenuation. The satellites at this orbital altitude are generally transferred to a graveyard orbit at least 300 km higher than the effective GEO at the end of their life. The large fragments in GEO are relatively easy to remove because most of the GEO fragments are at roughly the same orbital altitude, and the velocity variation required to remove multiple objects is small. However, the removal of small debris in GEO is more difficult than in LEO because the GEO debris smaller than 50 cm is extremely difficult to be tracked on the ground. The method of lowering the orbital altitude of GEO debris and then sending it into the Earth's atmosphere is also feasible, but will require a big change in velocity. If the orbit of GEO debris is lowered through propellant consumption, the service life of the mission spacecraft may be sacrificed and multiple descents can not be achieved. Therefore, only the continuous low-thrust technologies (such as ion engine or solar sail) can be used generally. However, the GEO debris to be removed by the Earth's atmosphere needs to enter the LEO, which increases the chance of colliding with other objects.

Based on the above analysis, the laser push-off technology can be used to remove the LEO debris without generating new debris and can be reused, but is difficult to realize due to the limited technology level. In addition, the cost of this removal is relatively high. In contrast, the container collection method faces many technical problems in power control and in the avoidance of other spacecrafts, and the collector itself may be damaged to generate secondary debris, but the cost of collection method is very low. The capture deorbiting methods (such as space tug and tether) have the advantage of removing large-sized space debris in different orbit ranges, but have some disadvantages as well. For example, such rigid-flexible coupled structures involve complex key technologies, and have an operating range limited to the removal of space debris at close range.

### 7.4.4 Scheme Evaluation

In summary, each of the above technical schemes has its own disadvantages and advantages, so these removal schemes should be comprehensively evaluated. Considering different mission scenarios and needs, a set of comprehensive evaluation indexes need to be designed, mainly including three categories: adaptability, affordability, and practicability (or implementability). The affordability indexes can be further divided into four sub-categories: deployment affordability, operation affordability, R&D affordability, and manufacturing affordability.

The first index is adaptability, which is used primarily to measure the practicality and cost of removal options.

The second index is affordability. Among the various affordability indexes, the deployment affordability is mainly used to measure the system R&D costs; the operation affordability is used to measure annual operating and maintenance costs; the R&D affordability is used to measure the technology maturity; and the manufacturing affordability is used to measure the expenditures required to manufacture the debris removal systems.

The final index is practicability (or implementability), which is used to measure the actual efficiency and realization probability of a specific technology in a mission.

Based on the above index definitions, we can summarize the two schemes with relatively high comprehensive scores: electrodynamic tether [16] and ground-based laser. First of all, the advantages of electrodynamic tether are low operating cost and manufacturing affordability, as well as strong practicality. Because the tether is relatively simple, no additional propellant consumption is required during the orbital descent. However, the tether system is vulnerable to debris and has poor survivability, so its adaptability score is low. The ground-based laser is more affordable in operation and deployment and is easy to operate because of its low operating and maintenance costs relative to the initial investment. The ground-based laser can be deployed on a site with the required optical equipment and infrastructure, so its deployment cost is also relatively low. Its weakness lies in the manufacturing affordability and R&D affordability. Since not all the components of ground-based laser have been fully tested, more research and development efforts and considerable manufacturing funds are still needed.

### *7.4.5 General Information About On-Orbit Tests*

The electrodynamic tether has attracted the attention of many researchers because of its simple structure and no need for propellant consumption. As a very potential method of orbital descent, the tether has gone through many on-orbit tests in addition to a lot of basic theoretical research in international community. These tests have laid a practical foundation for the future practical application of electrodynamic tether. Some specific on-orbit tests are shown in Table 7.2.

As can be seen from Table 7.2, the on-orbit tests of electrodynamic tether that have been carried out since 1980 involve key technologies of this tether such as tether release, current generation, electron collection and tether stability test. However, the failure rate of on-orbit tests is still high on the whole, indicating that the electrodynamic tether research still needs to be improved in many aspects.

Figure 7.3 shows the satellite structure in NASA's "TSS-1" test program that was carried out in 1992. The on-orbit test was used to:

(1) Verify the deployment and retraction of tether system;
(2) Verify the attitude holding;
(3) Verify the power supply of TSS;
(4) Collect the information about spatial ionization environment;

**Table 7.2** On-orbit tests of electrodynamic tether[17]

| No | Year | Research organization | Program | Achievement | Test result |
|---|---|---|---|---|---|
| 1 | 1980 | NASA/ISAS | K-9 M-69 | Release the electrodynamic tether by 38m | Failure |
| 2 | 1981 | NASA/ISAS | S-520–2 | Release the tether by 65m | Failure |
| 3 | 1983 | NASA/ISAS | Charge-1 | Produce the maximum 1kV voltage and 80mA current | Success |
| 4 | 1985 | NASA/ISAS | Charge-2 | Receive the VLF electromagnetic waves | Success |
| 5 | 1989 | CSA | OEDIPUS-A | Measure the electric and magnetic fields in the ionosphere | Success |
| 6 | 1992 | NASA/ASI | TSS-1 | Verify the stability of the tether under gravity gradient | Failure |
| 7 | 1993 | NASA | PMG | Verify the ion contactor technology of hollow cathode | Success |
| 8 | 1995 | CSA | OEDIPUS-C | Study natural and man-made waves in the ionosphere | Success |
| 9 | 1996 | NASA | TSS-1R | Obtain the ampere-level current in tether | Failure |
| 10 | 2003 | NASA | ProSEDS | None | Cancellation |
| 11 | 2010 | ISAS/JAXA | T-Rex | Verify the electron collection technology of electric bare tether | Success |

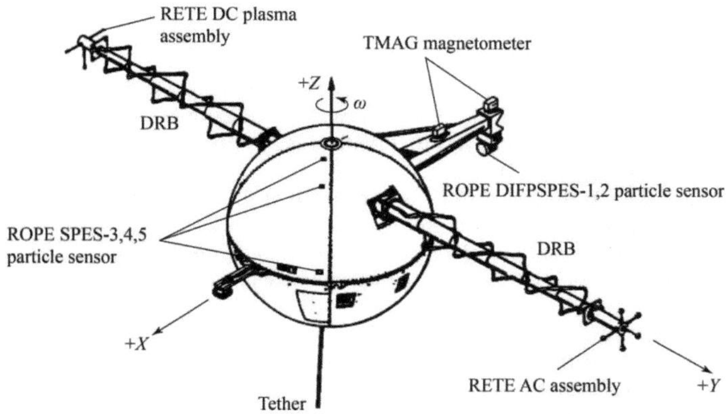


**Fig. 7.3** TSS-1 satellite structure[18]

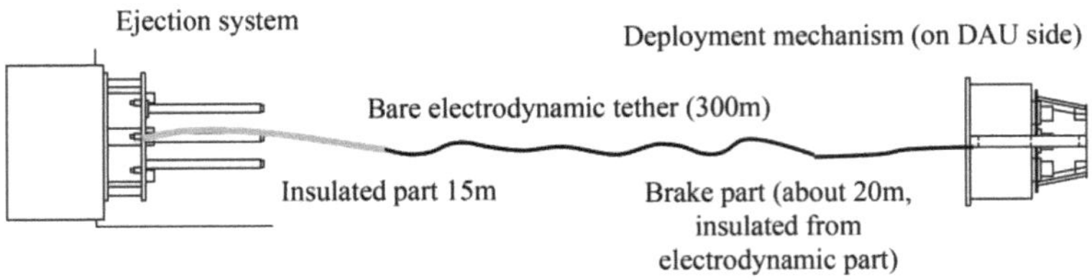


**Fig. 7.4** T-Rex satellite structure (see the colored illustration) (Fujii H, Watanabe T, Sahara H, et al. Space demonstration of bare electrodynamic tape-tether technology on the sounding rocket S520-25. AIAA Guidanc, Navigation, and Control Conference, 2011)

(5) Verify the control law of tether system; and
(6) Capture the satellites.

However, some of the tests were not completed on orbit because the tether could only be released for 256 m. 12 tests for collecting information on the ionospheric environment were completed during this period.

In order to complete the on-orbit tests of TSS-1, the NASA conducted the TSS-1R program in 1996 [19]. In the testing process, the electrodynamic tether was successfully released to a length of 19.7 km, but was suddenly broken on the side near the deployment device, leading to the failure of the program. However, the tests such as current generation, measurement of tether current, and measurement of the charge distribution around the charged satellite were completed.

The most recent successful test program was Japan's T-Rex program, as shown in Fig. 7.4. The purpose of the on-orbit test program was to verify the ability of a bare electrodynamic tether to collect electrons. During testing, the tether with a total length of 300 m was released at 3 m/s to a final length of 132.6 m.

## References

1. Tang, Q., Zhang, F., & Feng, Y. (2019). Analysis of advantages and disadvantages of space debris removal by electrodynamic tether. *Space Debris Research, 19*(2), 39–43.
2. Gong, Z., Zhao, Q., Li, M., et al. (2019). Frontiers and prospects of debris protection research. *Space Debris Research, 19*(3), 2–13.
3. Li, X., Yang, D., & Hou, T. (2017). Space Debris in Low Earth Orbit and its Hazards to the International Space Station. In: *Aerospace Safety Parallel System Committee of Chinese Institute of Command and Control. Proceedings of the 2nd China Aerospace Safety Conference* (Vol. 7). Aerospace Safety Parallel System Committee of Chinese Institute of Command and Control.
4. Chen, L., Zhang, W., Li, X., et al. (2020). Evaluation of the safety, reliability, maintainability and life of NASA international space station. *Quality and Reliability, 3,* 40–44.
5. Yan, N., Tang, Q., Chen, R., & Li, Y. (2018) Development of space debris removal technology in Europe and its implications. *Space Debris Research, 18*(2), 14–22.
6. Kessler, D. J., & Anz-Meador, P. D. (2001) Critical number of spacecraft in low earth orbit: using satellite fragmentation data to evaluate the stability of the orbital debris environment. *Space Debris, 473,* 265–272.

7. Yujun, Z., & Shuxing, F. (2010). Research on active debris removal. *Journal of the Academy of Equipment Command & Technology, 21*(6), 78–82.
8. Hakima, H., & Emami, M. R. (2018). Assessment of active methods for removal of LEO Debris. *Acta Astronautica, 144*, 225–243.
9. Emanuelli, M., Prasad, D., & Federico, G. (2014). Conceptualizing an economically, legally, and politically viable active debris removal option. *Acta Astronautica, 104*(1), 197–205.
10. Klinkrad, H. (2006). Space debris: models and risk analysis. *International Journal of Clothing Science & Technology, 18*(5), 314–325.
11. Han, Z., & Pang, B. (2012). Recent progress in space debris protection research. *Spacecraft Environment Engineering, 29*(4), 369–378.
12. Wang, W., Li, Y., Luo, W., et al. (2019). Analysis of space-based debris removal mission using laser. *Systems Engineering and Electronics, 41*(6), 1374–1382.
13. Wang, L., Lu, Y., Fang, G., et al. (2020). Overview and prospect of spacecraft drag-increase deorbiting technology. *Spacecraft Engineering, 29*(1), 61–69.
14. Li, R., Xiang, K., & Chen, Y. (2019). Analysis of ejection error in space debris removal by foam method. *Modern Defense Technology, 47*(2), 12–16+47.
15. Liu, H., Liu, D., & Jiang, Z. (2021). Overview and prospect of space manipulator technology. *Acta Aeronautica et Astronautica Sinica, 42*(1), 33–46.
16. Sanmartin, J. R., Lorenzini, E. C., & Martinez-Sanchez, M. (2010). Electrodynamic tether applications and constraints. *Journal of Spacecraft and Rockets, 47*(3), 442–456.
17. Yu, B., Wen, H., Jin, D., et al. (2016). Theoretical and experimental study on space electrodynamic tether system. *Advances in Mechanics, 46*, 226–266.
18. Dobrowolny, M., & Stone, N. H. (1994). A technical overview of TSS-1: The first tethered-satellite system mission. *IL Nuovo Cimento C, 17*(1), 1–12.
19. Stone, N. H., Raitt, W. J., & Wright, K. H., Jr. (1999). The TSS-1R electrodynamic tether experiment: scientific and technological results. *Advances in Space Research, 24*(8), 1037–1045.

# Chapter 8
# Technologies of Active Debris Removal

## 8.1 Introduction

Four measures are usually taken to deal with the space debris, including: avoidance (to avoid the impact of large cataloged debris), protection (to withstand the impact of numerous small debris), mitigation (to prevent the generation of new debris), and removal (to treat the debris that has already been formed). The first three measures can help improve the safety of space activities and prevent the generation of new debris to a certain extent, but have very limited or even no effect on completely controlling and reducing the formed debris and cleaning the space environment. In particular, the studies have shown [1] that even if no new spacecrafts are launched into space, the speed of debris generation by collision between space objects will be faster than the speed of natural debris removal by atmospheric drag. Therefore, the prevention measure alone is not enough to maintain the long-term stability of space debris environment. Active debris removal (ADR) is necessary for removing the existing space debris in orbit and thoroughly managing and protecting the space environment.

The debris removal activities usually focus on the orbits with high debris density (such as LEO or GEO), and provide the corresponding removal techniques for space debris in different size ranges. In Table 8.1, the proposed main ADR techniques [2, 3] are summarized and their application prospects are predicted. As can be seen from the table, different removal techniques have their own targets, and any of the techniques cannot remove all the debris in all orbits and all size ranges. Therefore, the removal technique selected varies with the mission objective.

So far, many removal techniques have been proposed, but most of them are still in the stage of conceptual exploration or technical research. Even though some techniques (such as space tug) have been demonstrated and tested, they are aimed at more large-sized cooperative objects than non-cooperative objects and small and medium-sized debris. In addition, the space community has not yet found a technically and economically effective removal program. To manage the space debris, a variety of



J. Zhang et al., *Space Debris Research*, https://doi.org/10.1007/978-981-96-9570-6_8

**Table 8.1** Main debris removal techniques

| Removal technique | | Orbital altitude range | Objects to be removed | Application prospect |
|---|---|---|---|---|
| Push-off deorbiting | Laser | LEO (ground-based, air-based) LEO, MEO, GEO (space-based) | 1–10 cm | High |
| | Ion beam | LEO, MEO, GEO | > 20 cm | High |
| | Solar sail | MEO, GEO | > 10 cm | High |
| Drag-increase deorbiting | Expansion foam | LEO | > 10 cm | High |
| | Electrostatic force | LEO | < (1–10) mm | Average |
| | Dust | LEO | < 10 cm | Low |
| Capture deorbiting | Space tug | LEO, MEO, GEO | > (0.1–1) m | Higher |
| | Space tether | LEO, MEO, GEO | > (0.1–1) m | Higher |
| | Container-like collector | LEO, MEO, GEO | < (1–10) mm | Average |
| | Space-based magnetic field generator | LEO, MEO | < 1 mm | Low |
| Autonomous deorbiting | Inflatable structure | LEO | EOL spacecrafts | Higher |
| | Brake sail | LEO | | Higher |
| | Solar sail | MEO, GEO | | Higher |

removal methods should be developed and applied comprehensively to form a cost-effective and complementary comprehensive management program. Next, each of the four main ADR techniques will be presented.

## 8.2 Push-Off Deorbiting Techniques

The push-off deorbiting is to use the interaction between the energy beam (such as ion beam, laser or solar radiation) generated by the mission spacecraft and the space debris to produce a specific force on the moving debris so as to force it off the original orbit and finally remove it. The push-off methods proposed by far include: ion beam method, laser method, and solar sail method.

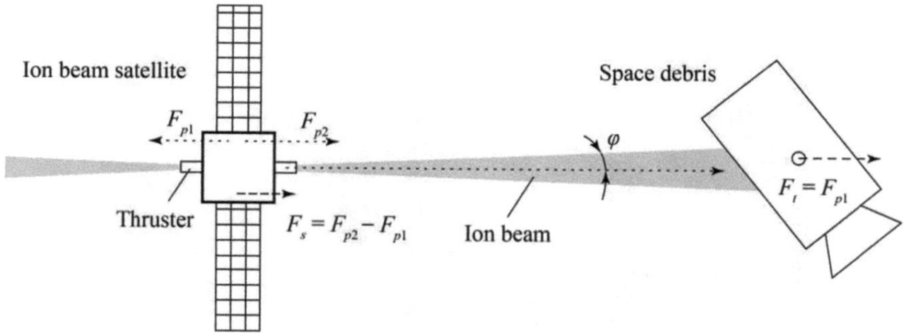


**Fig. 8.1** Working mechanism of ion beam (Kitamura S. Large Space Debris Reorbiter Using Ion Beam Irradiation. 61st International Astronautical Congress, 2010)

### *8.2.1 Ion Beam Method*

Compared with the contact-type deorbiting technology, the ion beam method has attracted more and more attention from the space community because it does not require direct contact with non-cooperative debris and has potential advantages such as simple operation, long service life, safe operation and wide applications [4]. This method is applicable to medium-sized and slightly large-sized debris, but its efficiency and economy are lower than other push-off removal methods.

The concept of using ion beam for attitude and orbit non-contact control, which was first proposed by Bombardelli and Peláez [5], can also be effectively transferred and applied to the non-contact deorbiting of space debris either in LEO,[1] GEO or GTO [6]. Its rationale is that the mission spacecraft ejects ion beams to continuously track and push the space debris so as to continuously increase/decrease its orbital tangential velocity and raise/lower its orbit (Fig. 8.1).

The foreign studies on the removal of orbital junk by ion beam jet were mainly led by Professor Claudio's team in Spain. Claudio proposed the concept of "Shepherd" satellite, which consisted of two reverse-mounted ion thrusters. The primary thruster ejects electrically neutral high-speed ion beams to space debris. The ion beams hit the space debris and exchange momentum with it, applying a drag to the debris in LEO (or a thrust to the debris in GEO) to slow it down and lower its orbit (or speed it up and raise it to the graveyard orbit). The secondary thruster needs to eject the ion beams at the same time, giving the satellite the same acceleration as the debris to keep tracking and pushing the debris. The team also studied and described the dynamics of debris removal by ion beam satellite [7, 8].

In 1959, the first ion thruster was developed by NASA's Harold R. Kaufman [9]. In 1964, the satellite carrying a mercury-ion thruster and a cesium-ion thruster was launched, which was the first on-orbit test of ion thruster. Since the 1980s, xenon has

[1] Liedahl D A, Libby S B, Rubenchik A. Momentum transfer by laser ablation of irregularly shaped space debris [C]. AIP Conference Proceedings, American Institute of Physics, 2010.

**Fig. 8.2** Xenon ion thruster on DS1 (see the colored illustration)

been widely used as propellant in the world.[2] The ionization chamber with ring-tip magnetic field structure has also been developed. These improvements have greatly enhanced the performance of ion thruster, extended its service life, and pushed it gradually into the application stage.

In 1988–1993, the NASA carried out the "NASA SEP Technology Applications Readiness" (NSTAR)[3] program based on solar electric propulsion to develop an ion thruster with superior performance and longer life, while developing the directly driven solar electric propulsion system and micro modular electric propulsion system. The ion thruster developed in the program (Fig. 8.2) was successfully applied to the "Deep Space 1" (DS1) spacecraft [10] as its main propulsion system, demonstrating the feasibility of using ion thruster as a propulsion system through on-orbit tests. The ESA also launched the "ARTEMIS" satellite[4] with four ion thrusters in 2001 to keep the satellite attitude stable. However, due to the unsuccessful launch, the satellite did not enter the target orbit. Finally, the satellite could only be pushed by the ion thrusters into its target orbit after nearly 10 months under the low thrust. After that, Japan and the United States successively launched the spacecrafts with ion thrusters [11].

The experiments and applications of ion thrusters were accompanied by the exploration of relevant theoretical derivation and mathematical simulation. Since 1965, scholars have made a large number of studies on plasma distribution (such as density, potential, electron temperature, etc.) and ion plume distribution law [12–15]. In

---

[2] Bai Shu, Yang Jianbin, Yan Changlin, et al. Development of Large-diameter Xenon Cryogenic Pump [C]. Abstract book of "2014 Annual Conference of Chinese Vacuum Society", 2014.

[3] Henry M D, Brinza D E, Mactutis A T, et al. NSTAR diagnostic package architecture and Deep Space One spacecraft event detection [C]. Aerospace Conference, 2000.

[4] Killinger R, Kukies R, Holtz L V, et al. Orbit Raising with Ion Propulsion on ESA's ARTEMIS Satellite [C]. AIAA/ASME/SAE/ASEE Joint Propulsion Conference & Exhibit, 2002.

China, there are also some theoretical studies on the plume field of ion beams[16–18], but few researches on the method of push-off deorbiting using ion beams. At present, the debris removal method based on ion beam jets still faces many difficulties and problems [19], for example, the development of reliable ion thrusters with superior performance, the accurate modeling of ion jet plumes, the orbital maneuvering of ion beam satellites, the high-precision tracking of targets, the rapid identification of a target's motion parameters and shape parameters, the selection of optimal ejection application points, and the determination of optimal ejection parameters.

In general, the ion beam method is of great research value, but still faces the following problems to be solved: the ion beams are freely dispersed and are difficult to control; the ions reflected back from the object surface may pollute the sensitive parts of the tug; the arc effect generated under coulomb force may damage the solar components and electronic devices; the application of this method is limited because the fragment is required to carry an electromagnetic device to meet the requirement of electromagnetic force [20]. Therefore, the technology of push-off deorbiting using ion beams still faces many challenges before being applied in engineering.

### 8.2.2 Laser Method

The laser-aided removal methods are generally divided into two methods: burning and pushing. Burning is the use of the powerful continuous-wave laser with high energy to irradiate the debris, raise its temperature and even sublimate it. Pushing is the use of high-energy pulsed laser beams to irradiate the surface of the debris and provide a certain velocity increment for the debris to reduce its perigee altitude, so that the debris will be exposed to a greater atmospheric drag when passing the perigee and will have its orbital life shortened, as shown in Fig. 8.3. In addition, the laser-aided removal method needs to accurately detect, locate and track the target in advance, which is difficult to realize for the centimeter-sized or even smaller space debris. Therefore, this technology is more suitable for the removal of larger debris.

To burn and sublimate the debris, the energy density of high-energy laser arriving at the space debris should be large enough for the temperature of debris to rise sharply in a short time to the melting point or even boiling point of debris material, so that the debris can be melted and vaporized and finally be eliminated. The vaporization and evaporation threshold of common debris materials is generally within the intensity range of 0.1–1 MW/cm$^2$. Due to the limitation of transmitting power, the energy density irradiated by a single set of high energy laser to space debris is limited. Therefore, multiple sets of high energy lasers need to work together to accumulate enough energy density in a short time to burn and sublimate the debris [21]. Due to the high energy requirement and high hazard of direct debris burning, the laser-aided removal method is preferred.

According to the different platforms where the lasers are located, the laser-aided removal systems can be divided into three types: ground-based system, air-based

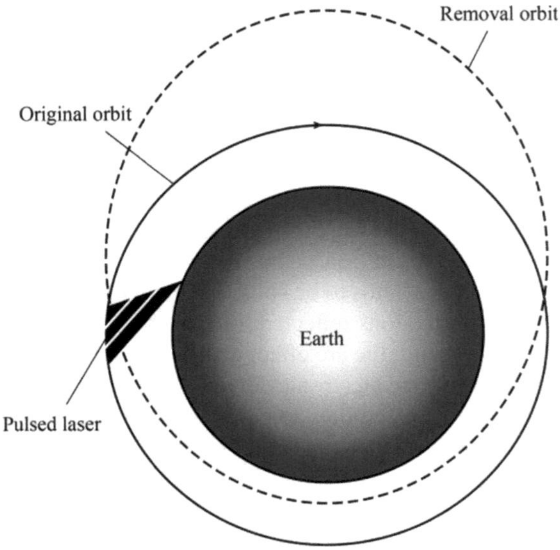


**Fig. 8.3** Schematic diagram of debris deorbit and removal by laser pushing

system and space-based system. Ground-based laser is more powerful, technologically more mature and more economical and reliable, and is easier to maintain and improve. But its requirements for the accuracy of measurement and control systems are higher, and its energy loss is larger due to the absorption by atmosphere. Due to the effect of turbulence and other factors, the laser is refracted and scattered. Due to the cloud shading, the laser system cannot work all time. Due to the influence of geographical location and distance, the spatial application range of the system is limited. In contrast, the space-based laser can propagate in vacuum with negligible loss and without the propagation error caused by refraction and scattering. With the change of platform deployment, its application range in space can be expanded. However, the space-based laser system has low technology maturity, high injection cost, high laser requirements, low reliability and a low capacity of maintenance and improvement [22]. In Table 8.2, the space-based and ground-based debris removal options are compared in terms of technical difficulty and removal efficiency.[5] The performance of air-based option falls between them.

The ground-based laser system for debris removal generally consists of five subsystems, as shown in Fig. 8.4.

(1) Control communication subsystem: responsible for collecting and processing the information provided by other subsystems, coordinating the operation of various subsystems, and organizing the removal work in a planned way.
(2) High energy laser: the core of the system. Its main component is a laser generator, which is responsible for generating high-energy laser beams.

[5] Kaplan M. Survey of Space Debris Reduction Methods. AIAA Space Conference & Exposition [C]. Padena, California, 2009.

**Table 8.2** Comparison of space-based and ground-based debris removal options

| Removal method | Technical difficulty | | | | | | Removal efficiency | | | | | Cost |
|---|---|---|---|---|---|---|---|---|---|---|---|---|
| | Launch | In-space assembly | Energy source | Energy storage | Heat-dissipation difficulty | Maintenance difficulty | Application time | Atmospheric transmission | Tracking and capture FOV | Energy consumption | Reuse | |
| Space-based laser system | Large mass and high difficulty | Required | Ground delivery or solar energy | Large difficulty | High | High | Relatively long | Basically not affected | Large | Small | Yes | High |
| Ground-based laser system | Relay mirror | Not required | Nuclear energy | Not required | Low | Low | Short | Largely affected | small | Large | Yes | Low |

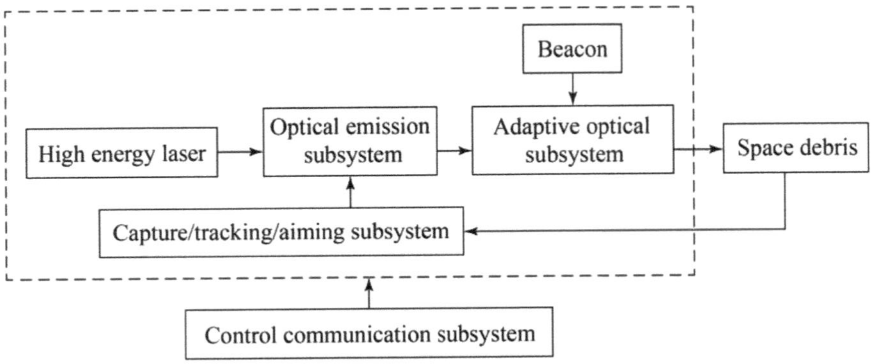


**Fig. 8.4** Components of ground-based laser system [23]

(3) Capture/tracking/aiming subsystem: the capture subsystem is responsible for distinguishing the space debris from the space background and detecting the general direction and dynamic parameters of the debris. The tracking subsystem is responsible for enabling the line of sight (LOS) of the instrument to follow the motion of the debris. The aiming subsystem is responsible for aiming the optical axis of the emitter at the object and firing the laser to the aiming point on the object. In order to stably irradiate the laser beams on the fragment, the laser beams are required to accurately hit the fragment and have small spot jitters.
(4) Optical emission subsystem: responsible for sending laser beams into the far field and converging them on the space debris to form a spot with enough high power density. The capture, tracking and aiming techniques allow the transmitting telescope to always track and aim at the target, locking the laser beams onto the debris.
(5) Adaptive optical subsystem: responsible for correcting the atmosphere-caused laser beam distortion in the transmission channel from the transmitting telescope to the space debris, and focusing the laser energy on the debris. It includes a wavefront error sensor and a wavefront corrector.
(6) Beacon: responsible for measuring the atmosphere-caused laser wavefront distortion in the transmission channel from the transmitting system to the space debris. Light from the direction of the debris must be used as the information carrier of the wavefront distortion. In the case of non-cooperation, artificial beacon light is needed.

In general, the ground-based laser system for debris removal works in the following way. At first, the radar or optical detector detects a debris target and sends the information data about the debris to the control communication system. After confirming the target, the control communication system guides the precision capture/tracking/aiming system to capture and lock the target. Then the capture/tracking/aiming system guides the optical emission system to aim at the debris. When the debris is in place, the control communication system sends out a removal command and starts the laser, which, in turn, emits the light beams to get rid of the debris target.

Compared with the ground-based system, the space-based laser system can be deployed on different orbits, and can remove multiple fragments on the same orbit or different orbits through maneuvering/orbital transfer and attitude adjustment. It is not affected by atmospheric attenuation, and has a larger removal range.

The space-based laser system is generally composed of six subsystems, as shown in Fig. 8.5 [24].

(1) Dynamic subsystem: the main function of this subsystem is to provide the control force required for orbit control, attitude control, axial acceleration and braking of the space-based laser platform in space flight, so as to remove the space debris on different orbits.
(2) Control subsystem: Based on the power supply and distribution of the whole system, the subsystem outputs control instructions to the power subsystem and high energy laser through the control algorithm, coordinates the operation of other subsystems, and controls the debris removal task in a planned way.

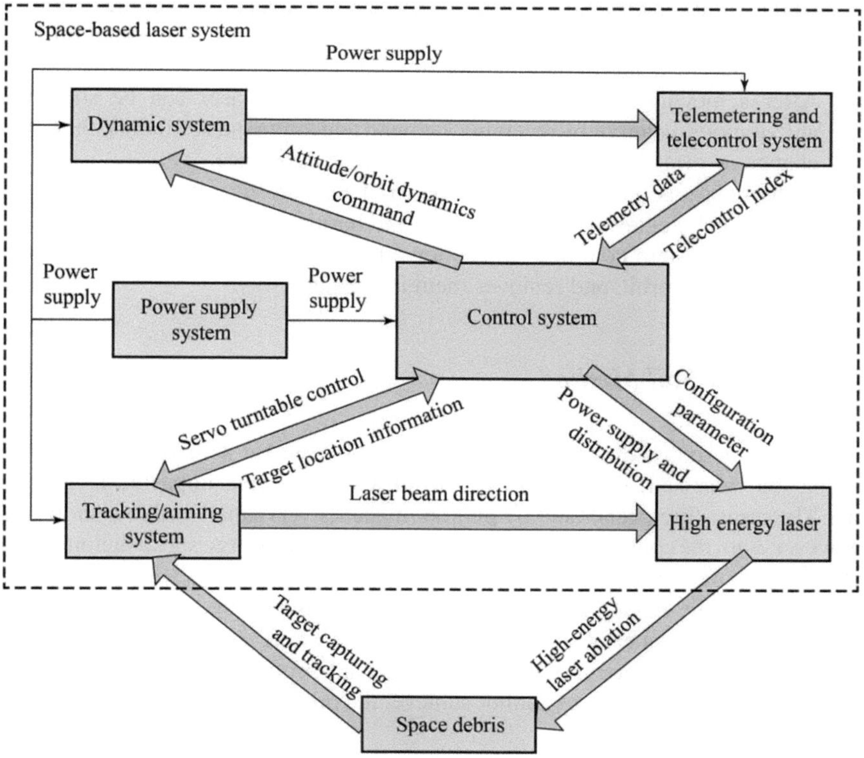


**Fig. 8.5** space-based laser system for debris removal [24] (see the colored illustration)

(3) Power subsystem: as the energy source of the whole system, the power subsystem ensures the normal operation of the whole system and provides the energy needed to remove space debris.
(4) Telemetering and telecontrol subsystem: the subsystem is used for the two-way information transmission between the space-based subsystem and the ground by transmitting the state information of the space-based subsystem to the ground and receiving the remote control command from the ground.
(5) Tracking/aiming subsystem: the subsystem can capture, identify, track and aim at the space debris, and converge high-energy laser beams to continuously irradiate the debris.
(6) High energy laser: the subsystem is the core of the whole system and can generate the high energy laser beams needed to remove space debris.

The specific process for debris removal by the space-based laser system based on the above subsystems is envisioned[6] as follows:

(1) At first, the detection device of the space-based platform finds a debris target within the removal range.
(2) The detection device sends the debris information to the control subsystem, which guides the tracking/aiming subsystem to track and aim at the debris.
(3) The telemetering and telecontrol subsystem sends the image and other information about the detected debris to the ground.
(4) After achieving accurate aiming, the high energy laser can be started by autonomous means or by receiving a ground telecommand to irradiate the debris, change its original orbit and reduce its altitude at perigee.
(5) If the debris is quite large, it can be irradiated on several consecutive orbits before finally entering the atmosphere and burning up.
(6) Then, the system continues to capture and track other fragments by adjusting its attitude or orbit, and removes them in the same way.

### *8.2.3 Solar Sail Method*

Sunlight is composed of the photons without static mass but with momentum. When hitting the solar sail, the photons can exert a force on the impacted area of the sail. The thrust force generated by photon impact is very small within a short time. However, with the continuous accumulation of this force, a large speed increment can be formed. In the space without air drag, the sunlight photons will continuously strike the sail, gradually raising its momentum to achieve the desired acceleration. Since there are a huge number of sunlight particles in the whole space, the solar sail using them as power source does not need to consume any propellant. With the advantages that other propulsion systems cannot achieve, the propulsion system based on solar

[6] Shen Zhenmin, Shang Weidong, Luo Pingping. Analysis of Laser-aided Debris Removal Technology[C]. Proceedings of the Second Symposium on Space Targets/Debris Monitoring and Removal Technology and Its Application, 2015.

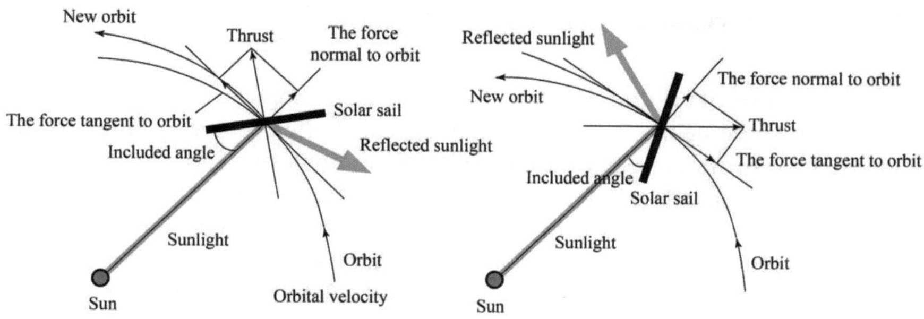


**Fig. 8.6** Using solar sail to push debris off orbit (see the colored illustration)

sail can be widely used in the solar system flight missions with low cost and large speed increment. It is a new space propulsion method with good prospects.

According to the basic theory on solar sail, the solar sail can also be used to push debris off orbit so as to complete the removal task, as shown in Fig. 8.6. If a fragment in an Earth orbit is loaded (in autonomous deorbiting) or attached (added through the mission spacecraft) with a solar sail, the solar sail can be pointed to the sun to generate a radiation pressure. Through continuous action, the radiation pressure can force the fragment off its original orbit, lift or lower it to a new orbit, and finally deorbit it. The trajectory of the debris can be raised or lowered by controlling the geometrical relationship between the sail and the sunlight. These operations are usually performed by the control gear on the sail, but a long operation time is needed to generate enough radiation pressure. A typical example of real-world solar sail is the UK's "CubeSail" program, in which a 3 kg NanoSat was launched to verify the technical feasibility of using solar sail for debris removal. The "CubeSail" can be attached to a fragment and fall with it into the atmosphere [25].

Due to the existence of solar radiation pressure in the orbits of various altitudes, the solar sail method is very widely used. In low orbits (such as LEO), the combined action of solar radiation pressure and atmospheric drag can be used to increase the drag, so as to accelerate the debris orbit descending. In high geosynchronous orbits, the solar radiation pressure becomes the main perturbation force or main deorbiting force to raise the debris orbit. J. Angel Borja et al. took a three-axis stabilized satellite in GSO as an example to study the technical feasibility of using solar radiation pressure to push the satellite off orbit, and mainly analyzed the influence of solar pressure on the satellite's semi-major axis and eccentricity [26].

At present, the main institutions engaged in the research of solar sails include Russia Babakin Space Centre, RSA, ESA, DLR, University of Glasgow, Goddard Space Flight Center, Jet Propulsion Laboratory, Langley Research Center and MSFC. Since 1993, Russia, the United States and Japan have carried out 8 solar sail experiments. Only Japan's "Ikaros" (with a mass of 315 kg), a Venus probe launched on May 20, 2010, successfully flew by Venus on December 8, 2010 [27].

## 8.3 Drag-Increase Deorbiting Techniques

The main function of the drag-increase deorbiting technology is to increase the flight resistance to space debris, reduce its flight speed, shorten its service life in orbit, and make it crash after re-entry into the atmosphere. In this technology, the satellite performing the removal task (the removal satellite) keeps a large distance from the space debris, thus avoiding direct contact with the debris and reducing the removal difficulty. Moreover, compared with natural drag (atmospheric drag), this technology has a more obvious effect on reducing the life of the debris. By increasing the area-mass ratio of the debris, the atmospheric drag to the debris can be increased. For the fragments of different sizes, different drag-increasing methods should be used to force them off orbit. In addition, this technology is more suitable for the removal of LEO fragments and is more effective in increasing the drag to them because they are exposed to greater atmospheric drag. The drag-increase deorbiting methods studied in this section are mainly applicable to the fragments that are unable to leave orbit by themselves and rely on the mission spacecraft to increase drag.

### 8.3.1 *Electrostatic Force*

The electrostatic force method takes advantage of the characteristics of space plasma. At first, the target debris is charged by the power supply of the mission spacecraft. Then the debris is exposed to the coulomb force in the space electric field, and then is slowed down and forced into a lower orbit, and finally is removed.

Since the potential of an object in LEO is almost equal to that in the space plasma environment, the debris must be provided with enough electric charges before being removed by electrostatic force. Usually, the interaction between an artificial charging grid and the space plasma can be utilized to inject negative charges into the debris.

The concept of using electrostatic force to increase the drag and remove the debris is shown in Fig. 8.7.[7] The electrodes of a debris-charging grid (bias grid) placed in LEO are charged by space plasma. The bias grid is a positive electrode relative to the space plasma environment. When approaching the bias grid at an average speed of 10 km/s, the debris is gradually charged with negative charges. Under the electrostatic force (for increasing the drag), the velocity of the debris gradually decreases. Then the orbital altitude of the decelerated debris decreases accordingly until the debris re-enters the atmosphere and crashes.

[7] Toyoda K, Cho M, Masui H, et al. Preliminary investigation of space debris removal method using electrostatic force in space plasma [C]. AIAA Aerospace Sciences Meeting Including the New Horizons Forum & Aerospace Exposition, 2009.

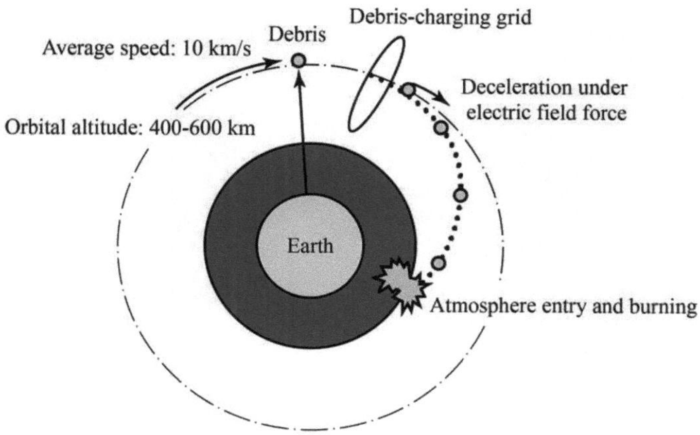


**Fig. 8.7** Concept of using electrostatic force for drag increasing and debris removal

### *8.3.2 Expanding Foam*

The Expanding Foam method also relies on the atmospheric drag to the foam ball (debris). Usually, it takes a very long time for space debris to fall naturally relying on atmospheric drag. However, if the mission spacecraft is used to spray foam to the debris and increase its area-mass ratio, the atmospheric drag to the debris can be significantly increased. Compared with natural attenuation, this method can significantly reduce the orbital lifetime of debris. The advantage of this method is that the direct contact between the mission spacecraft and the target debris is avoided so that the risk of colliding with the debris to be captured and removed is reduced. One of the key techniques used in the expanding foam method is the reliable foam expansion and debris adhesion in vacuum environment.

The expanding foam method is also inherently dependent on atmospheric drag, and thus is more suitable for removing the LEO debris, as shown in Fig. 8.8. At the same time, the mission spacecraft needs to be able to chase the target debris in order to create (or attach) spherical foams around/to the debris to increase its area-mass ratio. Then the atmospheric drag to the debris can be effectively increased and the debris's flight speed can be significantly reduced. In this way, the hundreds-year orbital life of the debris can be drastically shortened and the debris can re-enter the atmosphere within a specified time frame. The mission process for a spacecraft to remove debris using foam is composed of three main stages:

1. **Rendezvous stage**

After being launched, the mission spacecraft approaches the target fragments one by one in a predetermined order. The initial orbit of the spacecraft is assumed to be the orbit of the first target fragment, so the first orbital maneuver of the spacecraft after entering orbit is aimed to rendezvous with the target fragment.

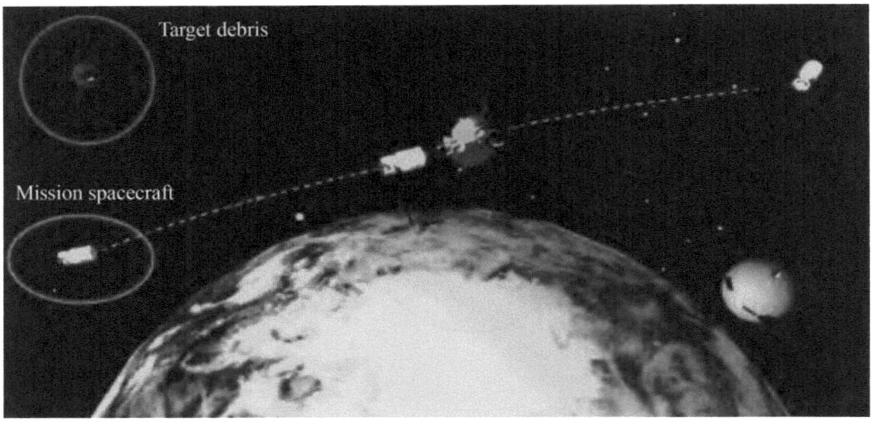


**Fig. 8.8** Concept of using expanding foam for debris removal

2. **Foaming stage**

After rendezvousing with the target fragment, the mission spacecraft enters the foaming stage and uses the foaming device to spray and attach foams to the fragment so as to increase its area-mass ratio and drag.

3. **Deorbiting stage**

After the spherical foams emitted by the mission spacecraft are attached to the fragment, the area-mass ratio and atmospheric drag of the fragment can increase. Then the fragment spirals down until it re-enters the atmosphere and crashes. After completing the current removal mission, the spacecraft can go to the removal of the next target fragment. Later in the mission, the spacecraft can also use thruster or the remaining foams to deorbit itself to prevent itself from becoming debris [28].

In addition to the significant advantage of not requiring a capture mechanism, the expanding foam method has the advantages of uncontrolled reentry, relatively stable foam production, and no potential hazards to ground system. This method cannot only remove large fragments, but also intercept small fragments and suck them into the foams or attach them to the foam surface. This is the so-called domino effect.[8]

In 2011, the ESA put forward the specific scheme of using expanding foams to increase the drag and deorbit the debris, as shown in Fig. 8.9. When the removing satellite rendezvouses with a fragment and flies around it, the foam sprayer installed on the satellite sprays foams to the fragment surface, which is then covered with foams and gradually becomes a foam ball. Due to the low density and large volume of the foam ball, the area-mass ratio of the fragment increases,[9] leading to the increase

[8] Andrenucci M, Pergola P, Ruggiero A, et al. Active Removal of Space Debris—Expanding foam application for active debris removal, 2011.

[9] Pergola P, Ruggiero A, Andrenucci M, et al. Expanding Foam Application for Active Space Debris Removal Systems [C]. 62nd International Astronautical Congress, 2011.

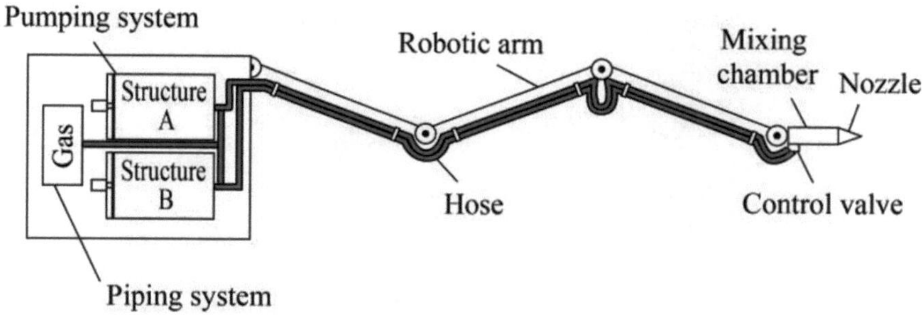


**Fig. 8.9** ESA's satellite using expanding foams for drag increasing and debris deorbiting

of aerodynamic drag that forces the fragment to leave orbit. To effectively remove the fragment, the foaming system often works together with electric propulsion system. When the fragment becomes a foam ball, the removing satellite uses electric propulsion system to deorbit the foam ball and accelerate its entry into the atmosphere. At the end of the mission, the satellite is also deorbited by electric propulsion system.[10]

### *8.3.3 Dust Method*

Different from the above two drag-increasing techniques, the dust method mainly relies on the force generated by the collision between debris and released dust particles to reduce the orbital altitude and life of space debris. However, atmospheric drag is still needed to assist the orbital descent, so as to remove the debris.

The physicists and engineers from the Plasma Physics Branch of Naval Research Laboratory (NRL) and the Naval Space Technology Center are working on a dust-aided drag increasing technology that can actively remove the LEO debris,[11], 2012-06-20 as shown in Fig. 8.10. The idea is to use a rocket to drop tiny bits of man-made dust (tungsten powder is considered) over the path of a fragment. When the fragment and the dust meet, the collision between them will generate a drag, causing the decrease of fragment orbit altitude, the increase of atmospheric drag and finally the acceleration and crash of the fragment. The debris expected to be removed is the tiny, dangerous and untraceable debris in LEO.

However, the released dust itself may become new space junk and pose a certain threat to the spacecraft in service. In addition, considering various factors such as launch capacity and cost, dust particle size and release density, debris size and debris density in space, this technology is still in the early stage of development and needs to be further studied.

---

[10] Pergola P, Ruggiero A, Andrenucci M, et al. Low thrust Missions for Expanding Foam Space Debris Removal [C]. International Electric Propulsion Conference, 2011.

[11] NRL scientists propose mitigation concept of LEO debris [EB/ OL]. https://www.nrl.navy.miV media/news-releases/2012/nrl-scientists-propose-mitigation-concept-of-leo-debris.

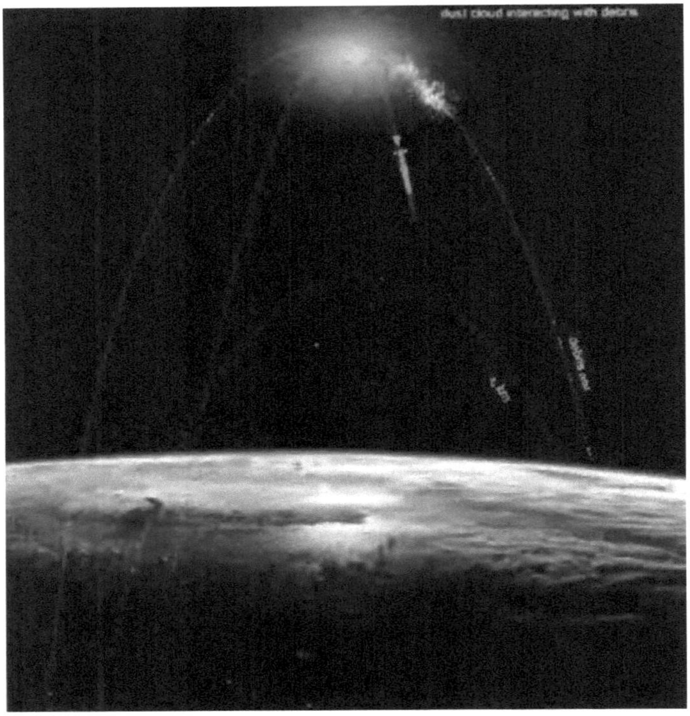

**Fig. 8.10** Concept of dust-aided debris removal

## 8.4 Techniques of Autonomous Deorbiting

For a defunct spacecraft (such as EOL satellite), the spacecraft will not become a piece of long-term orbiting debris if a set of deorbiting device is added to the spacecraft during its design and is activated near the end of the spacecraft's life to autonomously deorbit the spacecraft. The method of deorbiting the spacecraft by means of its own devices, as described above, is called autonomous deorbiting (space debris removal). The advantage of this method is that rather than relying on an additional mission spacecraft to assist the removal, the EOL spacecraft can be autonomously deorbited by itself, which greatly reduces the cost and design difficulty of the removal mission. In addition, the current debris mitigation measures have considered "using the reserved residual propellant to deorbit the spacecraft at EOL to reduce its residence time in space". The same deorbiting effect can be achieved if a self-deorbiting device without consuming propellant is added to the design of the spacecraft. At the same time, the fuel saved can extend the life of the spacecraft. The existing self-deorbiting devices which do not consume propellant mainly include brake sails, inflatable structures and solar sails.

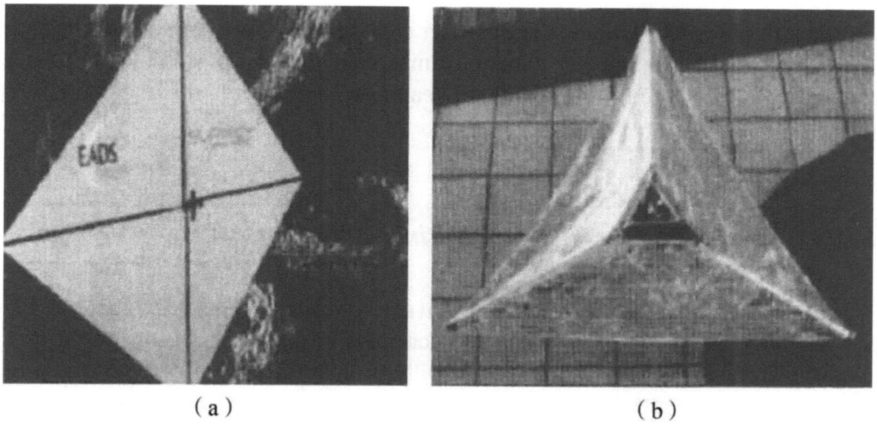

Fig. 8.11. Lightweight brake sails**a** Square; **b** Triangle

### 8.4.1 Brake Sail

Like the expanding foam method, the essence of brake sail method is also to increase the area-mass ratio of the debris so as to increase the atmospheric drag to the debris. In this method, various shapes of sails formed by spreading thin films on the satellite trusses are used to generate a braking drag that forces the satellites to leave orbit earlier. Generally, the brake sails have an area ranging from several square meters to dozens of square meters, and have different shapes, as shown in Fig. 8.11. With the advantages of small weight, simple structure and low cost, the brake sail is very suitable for small satellites in LEO.

### 8.4.2 Inflatable Structure

An inflatable structure can also be expand into a balloon, paraboloid or other shapes to improve the area-mass ratio of LEO spacecraft and finally increase the aerodynamic drag that forces the spacecraft to leave orbit earlier. In 2004, the American Bell developed a Towed Rigidizable Inflatable Structure (TRIS), which consisted of three brackets connected to a large parabolic device that was usually retracted in a small box. When the satellite was at EOL, the extendable brackets would be deployed and the inflatable device would be inflated into a paraboloid. The mass of a deorbiting device usually accounts for 1–1.6% of the total mass of the satellite. For 500–600 km LEO satellites, their deorbit time is 0.5–1 year [29].

In addition to the above parabolic device, spherical and polyhedral inflatable devices can also be designed. At the end of a satellite's life, the compressed nitrogen gas is released and a balloon is formed. If the mass of the satellite is 1200 kg, the orbital altitude is 830 km and the diameter of the balloon is 37 m, the satellite can leave orbit in one year and then enter the atmosphere to burn up. Without this device, the satellite will stay in orbit for 30–40 years. As can be seen, this method is very effective.

### 8.4.3 *Solar Sail (for Self-Deorbiting)*

If a spacecraft is designed with a solar sail, it can be self-deorbited at EOL by the solar sail and atmospheric drag. This technique can also be considered as one of the techniques of solar sail push-off deorbiting (with the solar sail on spacecraft) discussed in Sect. 8.2.3. The specific deorbiting process has been presented in Sect. 8.2.3 and will not be repeated here.

## 8.5 Techniques of Capture Deorbiting

The debris removal methods described in the previous sections are non-contact removal methods, which usually make use of the interaction between debris and microscopic or macroscopic media (such as photons or dust) to change the orbital velocity of the debris, which is then transferred to other orbits and finally removed. In those methods, the risk of collision between the debris and the mission spacecraft is reduced, but the operability is also heavily sacrificed. The opposite to them is the capture method that removes the debris by physical contact.[12]

According to the different physical contact modes, the spacecrafts used for debris capturing can be divided into four categories: space tugs, space tethers, space-based magnetic field generators and container-like collectors. The space tug and space tether, collectively called debris traps, are usually used to remove large fragments in a one-to-one manner. At present, the method of "capture + deorbit" is preferred, mainly using the techniques of non-cooperative targets capturing and efficient orbital transfer.

Space-based magnetic field generator and container-like collector are suitable for the removal of small debris. The former uses magnetic field to actively adsorb the surrounding small debris, while the latter relies on passive interception to capture the small debris below the centimeter level which is difficult to track accurately.

---

[12] Jared B, Stephanie C, Edward D, et al. Eliminating space debris: applied technology and policy prescriptions [R]. Global Innovation and Strategy Center, 2008.

### 8.5.1 Space Tug

A space tug usually consists of two parts: a robotic capturing arm (or manipulator) and a platform. The platform can autonomously rendezvous with, approach and dock at the debris. The platform is composed of a structural subsystem, a power subsystem, a thermal control subsystem, an integrated electronic subsystem, a telemetry, telecontrol and communication subsystem, a guidance, navigation and control (GNC) subsystem, and a propulsion subsystem. The manipulator can capture the target debris under autonomous control, and finally eliminate the debris through orbit maneuvering. Some techniques of the space tug have been proven in actual flight, but the cost of space tug is very high. Therefore, space tug is applicable to removing some urgent debris, such as large space debris.

As a rigid capturing mechanism, the space manipulator has the characteristics of high control precision and good capturing stability. Especially in the process of despinning and deorbiting a defunct spacecraft, the manipulator can accurately control the spacecraft. Therefore, the manipulator cannot only despin the defunct spacecraft, but also ensure its deorbiting reliability and safety.

The use of robotic arm for space operation first started with the "Lunakod/Luna" program of the Soviet Union. After that, Canada, the United States, Europe, Japan and other countries and regions successively carried out the research and development of space robots [30]. Some key technologies of the space tug used for debris removal have also been demonstrated in actual flight. The United States, Russia, ESA and other countries or organizations have developed a variety of spacecrafts with autonomous control and on-orbit rendezvous functions and with the ability to capture large debris. Among them, the Spacecraft for the Universal Modification of Orbits (SUMO) of the United States is the most representative space tug.

### 8.5.2 Space Tether

Since Konstantin Tsiolkovsky first described the tether system and its use in his book *Dreams about the Earth and the Sky* in 1895, researchers have continuously proposed the innovative ideas on tether application that made full use of the characteristics of tether (large flexibility and scale) to achieve unique mission objectives. After that, a wide range of tether systems were derived, including electric tethers, momentum exchange tethers, and towing tethers. As a flexible capturing method, the tether system has advantages such as small compression volume, small launch mass, long capture distance, small influence on the carrier, and good safety and flexibility compared with rigid capture methods (such as telescopic link and robotic arm). Therefore, tether is also considered to have broad application prospects.

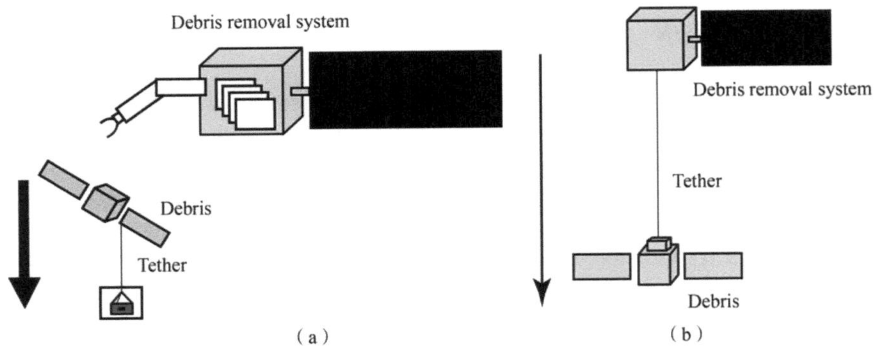


**Fig. 8.12** Two types of EDT **a** Terminator tether; **b** Reusable tether

1. **Electrodynamic tether**

There are two types of electrodynamic tether (EDT) (Fig. 8.12 [31]). One type is speed-increment EDT, which can be mounted on a space object before launch or attached to the object via a debris collector, but can only be used once. Therefore, 10 sets of such EDT system are typically carried by each space platform performing the debris removal mission. This EDT system is also called by some scholars as "terminator tether".[13] The other type is a reusable tether which is usually fixed to the debris collector platform on one end and is operated in a way similar to space tug. In general, the EDT is the connection device between the target and the satellite platform, and is also the dynamic device used for removing the debris.

The basic principle of space debris removal by EDT is shown in Fig. 8.13. When the conductive tether moves at orbital velocity in the geomagnetic field, the fast cutting of magnetic inductance lines by tether generates a strong induced electromotive force on the tether. As a result, the charged particles in the ionosphere are collected at the upper end of the tether and are emitted from the lower end of the tether to form a stable current. Meanwhile, the geomagnetic field produces a Lorentz force on the charged tether in the opposite direction of the movement velocity, thus reducing the orbital energy and altitude of the satellite [32].

The use of EDT can deorbit space debris much faster than the removal methods that rely solely on atmospheric drag. Compared with the capture and deorbiting methods that use chemical propellants to generate a propulsive force, the EDT has the advantages of high efficiency, low mass and no complex control system [33]. Therefore, EDT is theoretically a promising way of deorbiting the defunct LEO satellites.

[13] Hoyt R, Forward R. The terminator tether: autonomous deorbit of LEO spacecraft for space debris mitigation [C]. Proc. of 38th Aerospace Sciences Meeting and Exhibit, Reno Nevada, 2000.

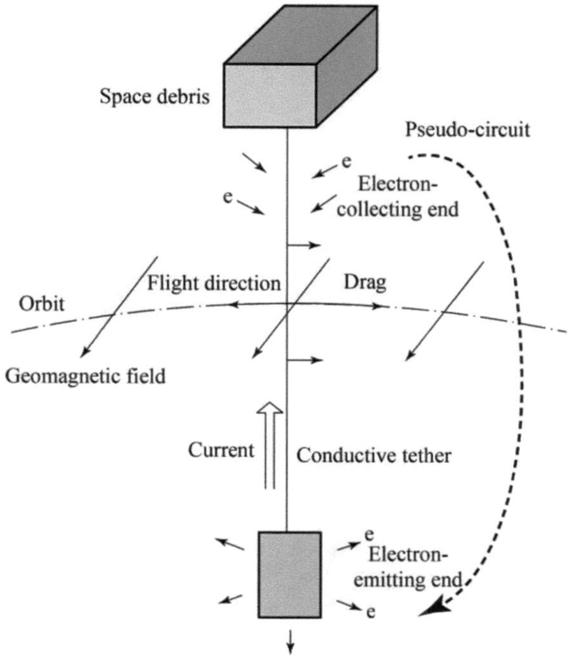


**Fig. 8.13** Operating principle of EDT [34]

2. **Momentum exchange tether**

Among the non-EDT satellite designs, the reusable Momentum Exchange Tether (MET) satellite system is more economical than conventional systems that consume chemical fuels. The basic idea of MET is to change the orbital velocity of space objects through the exchange of momentum between the surrounding objects. Its typical application is to raise the orbit of one satellite by lowering the orbit of another satellite. The momentum exchange between two satellites is usually "beneficial" to both satellites. For example, the momentum exchange between the pods of two space stations cannot only increase the stations' orbital altitude, but also enable the deorbit and reentry of cargo capsule. According to the motion of the tethered satellite relative to the orbital coordinate system, the tethered satellite systems can be divided into three types: pendent type, swinging type and rotating type. The most basic pendent MET is shown in Fig. 8.14. The pendent-tether satellite system consists of two satellites with different orbital altitudes, which orbit the Earth at the same orbital angular velocity as the center of gravity of the system. If the tether is cut at some point, the balance will be disrupted and the two satellites will enter their new orbits. When the tether is cut (the objects are separated from each other), the object in the higher orbit is traveling at a velocity faster than that necessary for maintaining a circular orbit. Therefore, its new orbit will be an elliptical orbit with perigee at the present position. The condition of the object in the lower orbit is exactly the opposite, so it will enter another elliptical orbit with apogee at the present position.

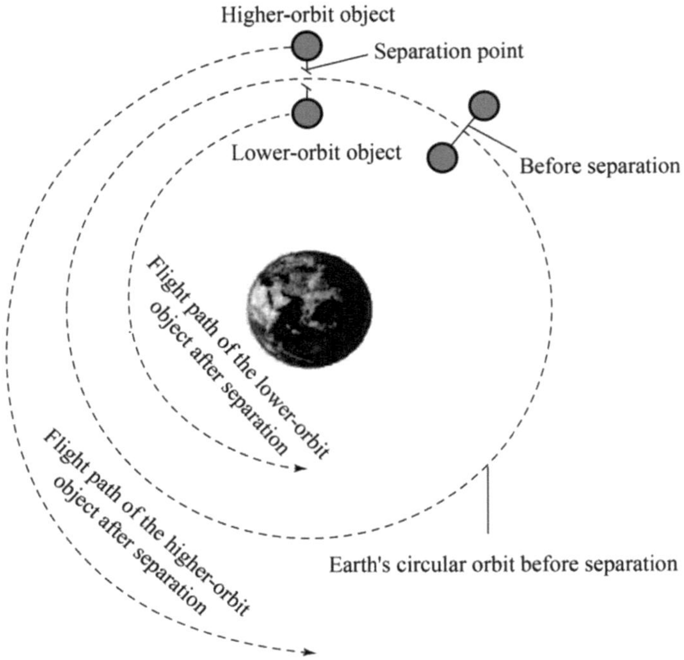


**Fig. 8.14** Pendent MET

The MET satellite system is usually not a simple pendant type, but a pendant type with moderate swinging or rotation to improve the efficiency of moment exchange. For example, when the tether swings or rotates in the same direction as orbital angular velocity, the speed of the satellite in the higher orbit will increase, while the speed of the other satellite will decrease or even reverse direction. Therefore, the MET satellites are very suitable for the debris removal missions that require orbital transfer.

3. **Towing tether**

Towing tether is a flexible tether that connects the mission spacecraft to the target object. In the removal mission, the mission spacecraft is usually powered to lower the orbit of LEO debris or pull the HEO debris to a higher orbit so as to achieve the purpose of debris removal. The tether only plays the role of transferring the towing force, and can be regarded as a flexible connection tether added to a space tug. Because the tether is flexible, its relaxation and intertwining must be avoided to prevent the collision between the debris and the tug.

### 8.5.3 Space-Based Magnetic Field Generator

The space-based magnetic field generator is used to generate a magnetic field with the aid of the mission spacecraft in order to attract the debris in orbit. Because the magnetic field intensity decreases rapidly with the increase of distance, the attraction of debris needs the generation of a large magnetic field, which puts forward harsh requirements for the spacecraft power system. The studies have shown that[14] the space-based magnetic field generators would require much higher voltage than that currently used in orbit in order to effectively remove debris. Even if the power system meets the requirements and generates a magnetic field strong enough to attract debris, the debris will still hit the mission spacecraft at high speed. As the debris approaches the spacecraft, the intensity of the magnetic field will gradually increase, and the acceleration effect on the debris will become more obvious. This requires a substantial increase in the protective structures of the spacecraft and in its weight and cost. At present, there are few studies on space-based magnetic field generator due to its low practicability, high cost and complex technology.

### 8.5.4 Container-Like Collector

The large-sized fragments pose a greater threat to the normal operation of a spacecraft, but are usually easier to track and observe so that an avoidance instruction can be sent to the spacecraft before collision. In contrast, a large number of small and medium-sized fragments widely distributed in space are often difficult to track, thus contributing to most of the space collisions. However, the capture device suitable for a single large fragment is usually not applicable to small and medium-sized fragments. One feasible solution is to build a passive collector that allows the fragment collision and then captures the fragments or slows them down and forces them off orbit.

So far, the proposed collectors mainly include aerogel collector or Large Area Debris Collector (LADC) and Multi-Layer Polymer Experiment collector (MUPLEX) [35, 36]. Next, the LADC is taken as an example to describe the main characteristics and progress of collectors.

As the main material of LADC, aerogel is a medium that traps small ultrahigh-speed particles. Its performance has been demonstrated in the laboratories. The impact dynamics, properties and effects, and nanostructural materials of aerogel have been explored for a long period with respect to the impact of cosmic dust. Therefore, LADC can be considered as a mature technique for capturing the micrometeoroids and small-sized debris within a certain range. However, the use of aerogel for capturing the medium-sized debris has not been studied. The collision between them is expected to produce secondary debris. Even so, the medium-sized debris

[14] Jared B, Stephanie C, Edward D, et al. Eliminating space debris applied technology and policy prescriptions [R]. Global Innovation and Strategy Center, 2008.

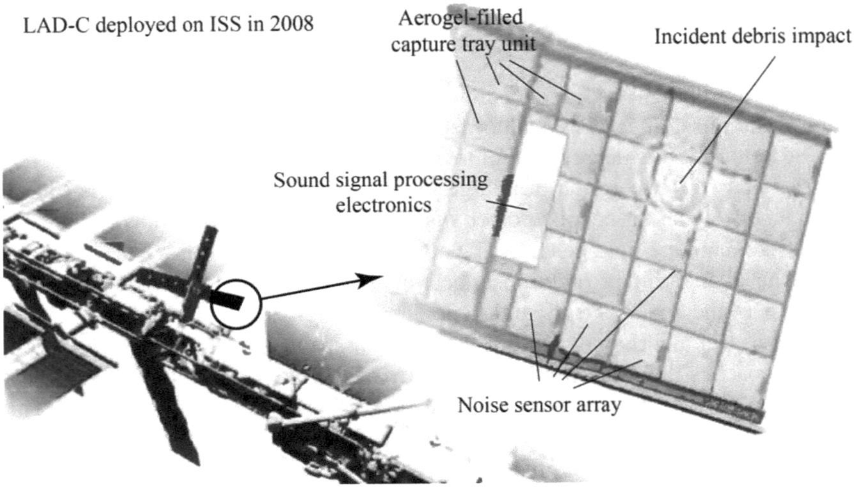


**Fig. 8.15** Schematic diagram of large-area debris collector

can be converted into small-sized debris through LADC so that the risk of collision between the debris and the spacecraft can be reduced.

The typical on-orbit flight test of container-like collector was the "LAD-C" developed by NRL. In 2002, the relevant scholars put forward the concept of on-orbit debris test. The NRL implemented this concept and, under the auspices of the US Department of Defense's "Space Test Program" (STP), eventually developed the "LAD-C", which was used in experiments on the International Space Station (ISS). The LAD-C consists of aerogel tiles and sound-sensitive elements. The former is used to capture the hypervelocity impact particles, while the latter is designed to record the impact characteristics of collected samples. The LAD-C is composed of the panels arranged in an inverted "T" shape, as shown in Fig. 8.15. Three arms are attached to the starboard frame S3 of the ISS. Each arm consists of four 1 m × 1 m × 2 m trays, each of which can be further divided into two aluminum panels. Each aluminum panel (6061-T6) consists of a number of small aerogel tiles, as well as 12 sound-sensitive elements attached to the aluminum frame. The aerogel collection area utilized on the ISS is about 10 $m^2$.

## References

1. Schfer, F., Lambert, M., Christiansen, E., et al. (2005). The inter-agency space debris coordination committee (IADC) protection manual. *Experimental Cell Research, 587*(587), 39.
2. Johnson, N. L. (2007). *Debris removal: An opportunity for cooperative research.* NASA Johnson Space Center.

3. Johnson, N. L., & Klinkrad, H. (2009). Space debris environment remediation concepts. *Orbital Debris Quarterly News, 13*(2), 5–6.
4. Xu, H. (2011). *Modeling and analysis of ground-based debris removal by laser*. Department of Aerospace Equipment, China Equipment Institute.
5. Lu, J., Ni, X., & He, A. (1996). *Physics of laser and material interaction*. China Machine Press.
6. Kitamura, S., Hayakawa, Y., & Kawamoto, S. (2014). A reorbiter for large GEO debris objects using ion beam irradiation. *Acta Astronautica, 94*(2), 725–735.
7. Bombardelli, C., & Peláez, J. (2011). Ion beam shepherd for asteroid deflection. *Journal of Guidance Control & Dynamics, 34*(4), 1270–1272.
8. Bombardelli, C., Urrutxua, H., Merino, M., et al. (2013). The ion beam shepherd: A new concept for asteroid deflection. *Acta Astronautica, 90*(1), 98–102.
9. Kaufman, H. R. (1990). Broad-beam ion sources. *Review of Scientific Instruments, 61*(61), 230–235.
10. Peng, Y., Sheng, Z., Zhang, H., et al. (2004). Mechanical behavior of space debris removal by high intensity laser. *Applied Laser, 24*(1), 24–26.
11. Kawaguchi, J., Uesugi, K., & Fujiwara, A. (2003). The MUSES-C mission for the sample and return—Its technology development status and readiness. *Acta Astronautica, 52*(2), 117–123.
12. Gyftopoulos, E. P., & Theodoridis, G. C. (1971). Cesium ion-beam neutralization with energetic electrons. *Aiaa Journal, 3*(8), 1414–1417.
13. Peng, X., Keefer, D., & Ruyten, W. M. (1992). Plasma particle simulation of electrostatic ion thrusters. *Journal of Propulsion and Power, 8*(2), 361–366.
14. Roy, R. I. S., Hastings, D. E., & Gastonis, N. A. (1996). Ion-thruster plume modeling for backflow contamination. *Journal of Spacecraft & Rockets, 33*(4), 525–534.
15. Katz, E., Willner, I., & Wang, J. (2010). Electroanalytical and bioelectro analytical systems based on metal and semiconductor nanoparticles. *Electroanalysis, 16*(1–2), 19–44.
16. Shang, S., Gu, Z., He, B., et al. (2015). Beam density distribution model for ion thruster. *Chinese Journal of Vacuum Science and Technology, 35*(12), 1414–1419.
17. Mao, G. (2009). *Spacecraft propulsion system and its application*. Northwestern Polytechnical University Press.
18. Zhang, Y. (2005). Research status, application and development trend of electric propulsion technology. *Journal of Rocket Propulsion, 31*(2), 27–36.
19. Cao, H., Chu, Y., Cao, Y., et al. (2014). Full particle numerical simulation of ion beam neutralization and propulsion in ion thruster. *High Voltage Technology, 40*(7), 219–2124.
20. Gong, Z., Xu, K., Mou, Y., et al. (2014). Space debris environment status and active removal techniques. *Spacecraft Environment Engineering, 31*(2), 129–138.
21. Qiang, X. (2000). Study on mechanical effect of high intensity laser interaction with target. *Laser & Infrared, 3*, 141–144.
22. Xu, H., Li, X., Li, Y., et al. (2011). Study on ground-based debris removal technology using laser. *Journal of the Academy of Equipment Command & Technology, 22*(3), 71–75.
23. Xu, H., Li, X., & Li, X. (2012). Research on orbital descent model of space debris irradiated by ground-based laser. *Modern Defense Technology, 40*(3), 18–23.
24. Hu, H., Zhang, Y., Yi, J., et al. (2014). Conception of space-based debris removal using laser. *International Space, 4*, 45–47.
25. Hou, Y. (2010). British research on the use of solar sail for orbital debris disposal. *Defense Science and Technology Information, 9*, 15.
26. Wang, W. (2007). Review of solar sail technology. *Space Recovery & Remote Sensing, 28*(2), 1–4.
27. Hang, G., Hong, X., & Kang, X. (2013). Status Quo and development trend of foreign space propulsion technologies. *Journal of Rocket Propulsion, 39*(5), 7-IS.
28. Liu, H., Liu, Y., Tan, C., et al. (2017). Analysis and suggestions on key technologies for space debris removal. *Spacecraft Engineering, 2*, 109–117.
29. Lin, L. (2012). Status Quo and removal of space debris. *Spacecraft Engineering, 21*(3), 1–10.
30. Zhang, X. (2009). *Research on the control strategy of space flexible manipulator*. Beijing University of Posts and Telecommunications.

31. Kawamoto, S., Makida, T., Saaaki, F., et al. (2006). Precise numerical simulations of electrodynamic tethers for an active debris removal system. *Acta Astronautica, 59*(1–5), 139–148.
32. Nishida, S., Kawamoto, S., & Okawa, Y. (2009). Space debris removal system using a small satellite. *Acta Astronautica, 65*(1–2), 95–102.
33. Kong, X., & Xu, D. (2010). A review of space tether research. *Spacecraft Environment Engineering, 27*(6), 775–783.
34. Wen, H. (2010). *Dynamic control of release and recovery of tethered satellite*. Institute of Vibration Engineering, Nanjing University of Aeronautics and Astronautics.
35. Liou, J. C., Giovane, F. J., Corsaro, R. D., et al. (2006). *LAD-C: A large area debris collector on the ISS*. NASA Johnson Space Center.
36. Kearsley, A. T., Graham, G. A., Burchell, M. J., et al. (2005). MULPEX: a compact multi-layered polymer foil collector for micrometeoroids and orbital debris. *Advances in Space Research, 35*(7), 1270–1281.

# Chapter 9
# Construction of Laws and Regulations on Space Debris

## 9.1 Construction of International Treaties and Regulations

### *9.1.1 Construction Process of Relevant International Laws and Regulations*

Since the launch of the first man-made Earth satellite in 1957, thousands of satellites and tens of thousands of cataloged man-made objects have been launched into orbit, which indicates that the human activities in outer space have gradually become normal. While benefiting human society, these activities have also brought a series of problems caused by space debris. Since the establishment of the Ad Hoc Committee on the Peaceful Uses of Outer Space in 1958, the United Nations has been committed to the study of space issues, the expansion and deepening of international cooperation among countries in the research and development of outer space, and the formulation and development of outer space law. In December 1959, the United Nations General Assembly adopted the Resolution 1471, transforming the previously established Ad Hoc Committee into a permanent body known as "the United Nations Committee on the Peaceful Uses of Outer Space" (UNCOPUOS), which was mainly responsible for reviewing the legal issues on outer space and drafting the outer space treaties. On December 13, 1963, the United Nations General Assembly adopted the *Declaration on Legal Principles Governing the Activities of States in the Exploration and use of Outer Space* (hereinafter referred to as the "Declaration on Outer Space"), which put forward several basic principles that must be abode by in the exploration and use of outer space, including the principle of "benefiting all mankind" reflecting the concept of "global symbiosis". Since then, the UNCOPUOS has formulated and adopted a series of relevant treaties, which constitute the basic system of outer space law, including five international treaties ("Outer Space Treaty", "Liability Convention", "Registration Convention", "Rescue Agreement" and "Moon Agreement"), as well as five non-binding documents called "principles" adopted by the



J. Zhang et al., *Space Debris Research*, https://doi.org/10.1007/978-981-96-9570-6_9

United Nations General Assembly. Among them, the "Outer Space Treaty" is the programmatic document of outer space law, while the "Liability Convention" and the "Registration Convention" are the main international legal documents concerning the pollution from space objects. These regulations provide a preliminary explanation of the concept of "space object", but do not give a clear definition, which has led to a lot of disagreement over whether the space debris is space object. Although the analysis of the provisions relating to space objects can serve as a theoretical guide to the space debris problem, there is still a lack of special treaties on the protection of outer space environment.

Except for the above-mentioned international treaties on the peaceful uses of outer space, there are only some guidelines and international documents on space debris mitigation. In 2002, the Inter-Agency Space Debris Coordination Committee (IADC), an intergovernmental organization formed by 11 countries, developed and adopted the IADC Space Debris Mitigation Guidelines ("IADC Mitigation Guidelines" for short), the first international document dedicated to space debris mitigation. In addition, the UNCOPUOS adopted the *UN Space Debris Mitigation Guidelines* ("UN Mitigation Guidelines" for short) submitted by the Space Debris Working Group of UNCOPUOS Science and Technology Group in 2007. The above two sets of international guidelines show that a set of regulations on space debris has been established by the international community, but its legal bond on member states is only based on the principle of voluntariness and is not as strong as an international law.

Although the international treaties on outer space activities are becoming more and more complete, there are only some guidelines on the protection of outer space environment. Those guidelines are only of advocacy significance and have no binding force of international treaties or agreements, thus playing only a limited role.

### 9.1.2 "Outer Space Charter"—Outer Space Treaty

The *Outer Space Treaty* is the core of the current legal regime concerning outer space and is also the legislative foundation of other outer space laws. It is a development and supplement to the *Outer Space Declaration*. This is the first time that the basic legal principles to be observed by all countries in outer space activities are set down in the form of an international agreement, so the Treaty is called "Outer Space Charter". The *Outer Space Treaty* stipulates the basic principles of outer space activities, the international liability for damage caused by space objects and the requirements for the protection of outer space and the Earth environment, which provide an important basis for the legal regulations on space debris. However, these principles are vague, providing the debris guidance based on scholars' interpretation of space objects and lacking specific implementation rules. Therefore, the Treaty is defective when being used as a legal guarantee for space debris regulations. In spite of the above problems (lack of relevant contents, vagueness of relevant concepts), the *Outer Space Treaty*, as the legislative basis of outer space law, is crucial for the formulation of legal regulations on space debris.

In 1967, the United Nations adopted the *Outer Space Treaty* (its full name is the Treaty on Principles Governing the Activities of States in the Exploration and Use of Outer Space, including the Moon and Other Celestial Bodies), also known as the "Outer Space Charter". The *Outer Space Treaty*, as the basis of outer space law, stipulates 10 basic principles to be followed in space activities. Among them, the principle of common interests, the principle of state responsibility and the principle of protecting space environment are closely related to the legal regulations on space debris pollution. As a core principle, the "principle of common interests" clarifies that the exploration and utilization of outer space "should be conducted in the interests of all countries" and that outer space is "the common domain of all mankind". It gives every country the right to explore and utilize outer space and the obligation not to put up any obstacles to other countries' free exploration and utilization of outer space. In addition, the Treaty also sets rules for international liability and environmental problems arising from space objects.

All the 17 articles of the *Outer Space Treaty* do not explicitly address the issue of space debris, but the articles 6, 7 and 9 potentially or indirectly deal with space debris. The articles 6 and 7 stipulate the international liability for damage caused by space objects and also embody the principle of compensation, providing a certain guiding role in the generation of space debris and in determining the liability for damage caused by space debris. According to the article 6, the States Parties shall be responsible for their own activities in outer space, including the moon and other celestial bodies, whether such activities are carried out by governments or non-governmental bodies (implying that private space activities may lead to State liability). According to the article 7, any State Party which launches or promotes the launch of an object into outer space, including the moon and other celestial bodies, and which uses its territory or equipment for the launch of an object, shall assume international liability when such object or its components cause damage to another State Party or its natural or legal persons on the Earth, in atmospheric space or in outer space, including the moon and other celestial bodies. This article requires that the State Parties conducting or facilitating the launch of a space object are jointly liable for damage arising from the space object. These two articles make clear that the States Parties are internationally liable for damage caused by their outer space activities, including the launch activities conducted by their private individuals and the launch activities assisting other States. In terms of space debris environment, the two articles can be applied to the attribution of liability for environmental pollution caused by space debris, and have expanded the liability of the States in order to maximize the interests of the victims. This regulation can indirectly promote Space States to regulate the outer space activities of their private entities and prevent and control the consequences of environmental pollution caused by those entities. The Article 9 deals with the protection of outer space environment. According to the article, in the research and exploration of outer space, including the moon and other celestial bodies, the States Parties shall avoid both harmful pollution to outer space and adverse changes that may result from the introduction of outer space materials into the Earth's environment, and shall take the necessary measures to control them. This indicates that, the international community has recognized that the human activities in outer space may bring about adverse effects on outer space and the Earth's environment, thus requiring the launching State to assume the responsibility of eliminating the harm. The article does not refer to space debris, but can be referred to and used as a basis in the study of international liability for environmental damage caused by space debris as a form of environmental hazard.

In terms of binding force, many articles of the *Outer Space Treaty* (such as the "principle of common interests", "principle of free exploration and utilization" and "principle of non-appropriation" stipulated in the first two articles of the Treaty) have been widely recognized by the international community and have the effect of customary international law. These articles apply to all members of the international community and are not limited to States Parties. In terms of operability, the Article 9 makes clear the importance of protecting the outer space environment, but does not specify how to implement it. The provisions of the Treaty are too vague and lack important implementation contents, resulting in the lack of a principle basis for the legislation on the control of space debris pollution. The *Outer Space Treaty* is the basis for the provisions of international outer space law that have been widely recognized by the international community. Although the term "space debris" does not appear directly in the *Outer Space Treaty* and a clear and detailed regulation on debris-related issues is absent from the Treaty, the Treaty has roughly constructed the origin of international law related to space debris. Of course, this construction is top-level and preliminary, because the relevant requirements are too rough and lack clear definitions and standards, and the definition of "space object" is so vague that some scholars have objections on whether the definition includes space debris. However, without the clear legislative basis provided by *Outer Space Treaty*, the legal construction related to space debris will become a water without sources or a tree without roots.

### *9.1.3 Liability Convention and Registration Convention*

The main international legal documents concerning the liability for space debris pollution include the Liability Convention and the Registration Convention, which refine and expand the relevant articles of the *Outer Space Treaty* and further develop the principles of international liability and liability for compensation given in the Treaty. To solve the problem of liability for the damage caused by space objects, the *Liability Convention* stipulates the system of liability for fault and the strict liability system, interprets the term "space object", and actually includes space debris into the scope of space objects. It clarifies that the objects in the concept of "damage" include persons and property, but fails to make clear the criterion of identifying the liability for fault. The *Registration Convention* mainly stipulates the compulsory registration obligation of State Parties when launching space objects, which is of certain positive significance for the prevention and control of space debris pollution. However, the connotation of "launching act" as the object of registration is relatively narrow, which makes it impossible to effectively prevent the debris pollution in advance. In addition, the Convention lacks the regulation on jurisdiction over space debris, making it difficult to identify the liability for space debris pollution.

In 1972, the 26th United Nations General Assembly adopted the *Convention on International Liability Caused by Space Objects* (hereinafter referred to as the Liability Convention), which has a total of 28 articles and is most closely related to the prevention and control of space debris pollution. Different from the *Outer Space Treaty*, which defines the subject of liability to pay compensation for damage caused by space objects in order to relieve the space environment pollution through damage compensation, the *Liability Convention* aims to prevent the occurrence of space environment pollution. From the "principle of State responsibility" and "principle of liability for compensation" established in the *Outer Space Convention*, the *Liability Convention* derives the fault liability system and strict liability system, that is, the fault liability applies to the space objects in outer space, while the strict liability applies to the space objects in the atmosphere. Specifically, when a space object of a launching State causes damage to a space object of another State, the launching State shall not be liable for the damage except that it is in fault. However, if the space object causes damage to persons or property in the atmosphere, the launching State shall be held liable whether it is in fault or not. If more than one launching States are involved, they shall bear joint liability. In addition, the *Liability Convention* also provides a series of procedural provisions on damages, such as setting the time limit of claims and establishing a claim settlement board as the dispute settlement mechanism. The definition of "space object" is further interpreted in the *Liability Convention*, which states that "space object, including its components and launcher and the components of its launcher", indicating that the space object includes its components and launch vehicle (and the components thereof). As the Convention ultimately fails to make clear what are the "components of space object" and whether the space debris belongs to such "components", it still lacks practical operability although it makes provision for the relevant party's liability for damages. However, according to the definition of space debris, the regulation of space debris pollution can be included in the scope of the regulating object of the *Liability Convention* by expanding the interpretation. At the same time, the Liability Convention defines the concept of "damage" as including the damage to and loss of persons and social property. It is of great significance for the establishment of an accountability mechanism related to space debris, as reflected in the following aspects:

(1) The liability for damages caused by space debris has been clarified. The *Liability Convention* regulates a liability for damages caused by space objects. Judging from the general definition of space debris, the space objects defined by the *Liability Convention* cover some types of space debris (e.g., failed spacecraft, and parts of the spacecraft). Therefore, on the basis of the *Outer Space Treaty*, the *Liability Convention* further clarifies that the States concerned should bear the liability for the damages caused by certain types of space debris.

(2) The principle of accountability for debris-caused damages has been established. In terms of the accountability mechanism, the *Liability Convention* defines two situations. First, the principle of strict liability is established for the damage caused by space objects to the Earth's surface and to the spacecraft in flight. Second, the damage occurred in outer space is defined as fault liability in Article 3 of the *Liability Convention*, that is, "Only when the damage is due to the fault of the previous State or its responsible person, the State shall be liable."
(3) A dispute settlement mechanism for debris-caused damages has been established. As for how to settle disputes after the occurrence of damaging acts, the Article 9 of the *Liability Convention* clearly states that "diplomatic channels shall be followed". If the diplomatic negotiation fails, the parties concerned shall establish a claim settlement board when any party makes a request (as first stipulated in Article 14), so as to establish a dispute settlement mechanism in which the board shall make a ruling.

The *Liability Convention* effectively fills the gap in the international mechanism of accountability for debris-caused damages, and is of landmark significance. This Convention lays a good foundation for the research of debris-caused compensation liability and dispute settlement and contributes to a large number of theoretical achievements in the space law field. However, the *Liability Convention* still faces huge problems in building a more stringent legal mechanism for space debris mitigation.

The *Liability Convention* stipulates that the launching State is liable for the damage caused by space object and that the basis of liability for the damage caused by space objects in outer space is the existence of fault. However, it does not provide clear criteria for the identification of a fault, which has caused a series of disputes in practice. Especially as the space debris management is put on the agenda, more and more problems are emerging due to the active removal of space debris. Take the collision between American and Russian satellites in 2009 as an example. Should Russia be deemed to be at fault for failing to remove the defunct Russian satellite (if considered as space debris) after it was disabled? But should the US have an obligation to plan the orbits of its satellites properly so that they could not collide with other countries' space objects? If such an obligation exists, the US should be equally at fault. Such questions are not answered by the *Liability Convention*. Therefore, the Convention still needs to be further refined in terms of the accountability for space debris pollution.

In 1976, the United Nations General Assembly adopted the *Convention on Registration of Objects Launched into Outer Space* (Registration Convention for short), which was a 12-article convention. The aim of *Registration Convention* is to establish a mandatory registration rule for space objects, namely, to establish a register at the United Nations and provide a means of identifying space objects. According to the Convention, when a State Party launches an outer space object, the object shall be registered at the United Nations. The establishment of the space object registration system aims to identify the source of space objects, so as to solve a series of problems such as the jurisdiction over space objects and the identification of the subject liable for the damages caused by those objects. For the damages caused by space objects whose source cannot be identified from the registered information, the other States with space monitoring capabilities should do their utmost to assist in the identification. The registration of space objects not only confers exclusive jurisdiction to the registrant State, but also satisfies the State's other substantive needs. The disclosure of launch information, object orbit and launch purpose, as well as the inherent relationship between space object and the specific subject of international law, can provide convenience for the formulation of space activity plans, prevent overcrowding in outer space and facilitate the adoption of collaborative relief measures. The registration of space objects can, to some extent, lock in the subject of an international legal liability that may arise from the launch of space objects, and facilitate international supervision of how the registrant States perform their obligations under international law in the space activities carried out by them or by their private entities. For the construction of relevant laws and regulations on space debris, the *Registration Convention* has indirectly enriched the relevant international laws for the rights/obligations organization mechanism and warning and prevention mechanism. With regard to the prevention and control of space debris pollution, the registration system established by the *Registration Convention* will record and announce the status of space objects. This will help prevent the occurrence of accidents such as collisions, so as to regulate space debris pollution from the perspective of beforehand preventing. The source/jurisdiction identification functions of the registration system are closely related to the active removal of space debris, and are helpful for solving the problem of space debris pollution at the level of after-event relief. However, in terms of beforehand preventing, the *Registration Convention* cannot play a significant role due to the lack of relevant contents. According to the Convention, the States Parties do not have the obligation to register the change of orbit parameters, the departure from a fixed orbit and the disintegration of space objects because these acts are not the "launch acts" stated in the Convention. However, the above three types of information are essential for preventing collisions between space objects. If the information cannot be disclosed to the international community in a timely manner, the damage-preventing function of the *Registration Convention* will be greatly weakened.

The *Registration Convention* formally fills the gap of the *Liability Convention* that lacks the rules for identifying the ownership of space objects, but still exhibits various loopholes in the actual implementation process. The fact that most of the launches with military intentions are not registered and that the Convention does not require the launching States to provide valid marks for identifying their spacecrafts and associated components makes it impossible to trace the ownership of relevant space debris according to the register prepared under the Convention. Another loophole in the Convention is that it does not give a clear definition of "space objects" to be registered—either including only artificial objects such as the satellites that are still in operation, or including not only the operational artificial objects but also other artificial objects in space (such as the defunct satellites, the satellite remains left in outer space after the mission failure, and the debris generated by explosions). From this point of view, the State Parties are not obliged to register with the United Nations the space debris produced in space activities, so the attempts to use the Convention to strengthen the tracking and disposing of space debris will be futile. Nevertheless, the *Registration Convention* is still of positive significance for the construction of legal mechanisms related to space debris, especially in the aspects of early warning and prevention.

### 9.1.4 IADC Mitigation Guidelines and UNCOPUOS Mitigation Guidelines

None of the above conventions with international legal binding force have directly addressed the issue of space debris. At present, there are only international guidelines and documents on space debris mitigation. Space debris mitigation is a relief measure to prevent space debris pollution in advance. Specifically, the measures such as avoiding the collisions between space objects and using safety protection technologies to avoid explosions are taken to reduce the possibility of space debris generation during the whole process of space object production, operation and disposal. The contents of space debris mitigation are mainly embodied in space debris mitigation technology, spacecraft protection technology, and monitoring and warning technology. On the basis of a clear definition of space debris, the *IADC Mitigation Guidelines* set out general guidelines and principles for debris mitigation programs, and put forward the implementable technical suggestions. With the deepening of practice, the experiences in the Guidelines are summarized and updated. On this basis, the *UNCOPUOS Mitigation Guidelines* set the minimum requirements for the protection of space debris from contamination.

In 2002, the *IADC Mitigation Guidelines* were formally adopted at the 20th IADC Conference. As the first specialized international document on space debris, the Guidelines required States Parties to formulate policies on space debris mitigation and fulfill their obligation of effectively controlling the proliferation of space debris. The debris mitigation plan proposed in the Guidelines has been highly recognized by the major space-faring countries and has strong technical authority, which is of great significance to the construction of laws and regulations on space debris and to the solving of space debris problem. The key elements of the *IADC Mitigation Guidelines* included: giving a clear definition of "space debris" as a non-functional man-made orbiting (or re-entry) object and its components and fragments; describing the up-to-date experience of space debris management, which was applicable to the whole life cycle of a spacecraft (from design, production, launch and operation, to subsequent disposal) in order to limit the disintegration of the spacecraft during its operation and the release of space debris; establishing general guidelines and principles for space debris mitigation programs, and recommending effective mitigation techniques, which included limiting the amount of operational debris released during normal launch activities and limiting the disintegration of space objects in orbit; and specifying special disposal principles for the spacecrafts at the end of their life so as to prevent in-orbit collisions. Finally, as the main content of the Guidelines, the concept of "protected area" was specially put forward to clarify the requirements for the disposal of LEO/GSO spacecrafts after the mission completion. Due to strong operability, the Guidelines have greatly promoted the international efforts in space debris control.

With the deepening of the research on debris mitigation techniques, the *IADC Mitigation Guidelines* were improved in 2007 to reflect technological advances and incorporated some of the up-to-date practical experiences. There were two major changes in the Guidelines. First, the general concept of "space system" was canceled and replaced by a refined concept of "spacecraft and upper stage". Second, to solve the problem of how to dispose the satellites at the end of their missions, two technical support conditions for keeping the satellites in GSO were clarified to control the phenomenon that non-functional satellites were staying in the orbital space to interfere with other spacecrafts.

In December 2007, the United Nations General Assembly adopted the *UNCOPUOS Mitigation Guidelines* submitted by the Space Debris Working Group of UNCOPUOS Science and Technology Group. The contents of the UNCOPUOS Guidelines are mainly based on the *IADC Mitigation Guidelines* and require member states to voluntarily implement the debris mitigation techniques and procedures through their respective mechanisms so as to meet the requirements of the Guidelines to the maximum practicable extent. As another special international document on space debris issues, the *UNCOPUOS Mitigation Guidelines* require member states to incorporate debris mitigation considerations into the launch preparations (including the design, manufacturing and mission planning) of spacecrafts and upper stages in order to limit the release of space debris and reduce operational debris in the planning and design stages. Meanwhile, to avoid the possibility of accidental collision between space objects in orbit, the member states are required to limit, to the maximum extent, the activities harmful to the outer space environment such as man-made self-destructive explosions, and restrict the long-term stay of spacecrafts in orbital space after the completion of their missions, so as to avoid interference with other spacecrafts.

In February 2008, the United Nations General Assembly (UNGA) adopted the non-binding *UNCOPUOS Mitigation Guidelines*, requiring member states to take the Guidelines seriously. In addition, the UNGA called on the international community to continue exploring and developing the ways to better track and dispose the space debris. The *UNCOPUOS Mitigation Guidelines* approved by UNGA have acquired the attributes of an international legal document. However, as specifically pointed out in the Guidelines, this document will not be tied to international outer space law at the current stage, so its rules do not have the mandatory binding force of international law and its implementation relies on the national mechanisms voluntarily formulated by member states. Therefore, although its level of effectiveness is higher than that of the *IADC Mitigation Guidelines*, this document also plays a limited role due to the lack of coercive power.

The most significant defect of either the *IADC Mitigation Guidelines* or the *UNCOPUOS Mitigation Guidelines* is that they do not have the binding force of international law. Thus, their implementation depends on national enforcement mechanisms. Since neither documents have the attributes of an international treaty or agreement, the development of national mechanisms depends entirely on the attitude of the States. Although the *IADC Mitigation Guidelines* have been implemented in many countries, their effect is limited due to their non-mandatory nature. Therefore, the UNCOPUOS Science and Technology Group added a new topic named "Long-term Sustainability of Outer Space Activities" to its agenda from 2010 in order to develop a code of conduct for maintaining the sustainability of outer space environment that could be widely accepted by the international community and implemented on a voluntary basis. Up to now, the *Rules on the Long-term Sustainability of Outer Space Activities* (draft) developed by UNCOPUOS are still in the negotiation stage as no consensus has been reached on some articles. The articles 20 and 22 of the Rules require the establishment of specific standards and procedures for the active removal of space debris, but are still under discussion as they involve the harmonization of national practice standards. The regulation on the active removal of space debris in the form of rules can lay a foundation for the formation of subsequent international laws and will be of far-reaching significance.

## 9.2 Construction of Relevant Laws and Regulations in Major Space Powers

### *9.2.1 Construction of Debris-Related Laws and Regulations in the United States*

As early as the middle of the 20th century, the United States has begun to explore and develop outer space activities. With a long history of space activities, the United States is the main producer of space debris, producing 45% of the total debris. For this reason, the United States attaches great importance to the space debris problem, trying to mitigate the space debris through technological research, policy support and legal regulation so as to protect the outer space environment and guaranteeing the development of its space industry.

To accelerate the development of space industry, the US Congress passed the *National Aeronautics and Space Act* in 1958, which has since been amended for several times. In the same year, the National Aeronautics and Space Administration (NASA) was created under the *National Aeronautics and Space Act*. The Act defined the functions of NASA, established the liability and relief system for damages caused by spacecraft, and specified the subject, scope, way and procedure of claim in detail. In the later decades, the US has led the world in space legislation.

In the space debris field, the NASA's Johnson Space Center and Marshall Space Flight Center have been studying space debris since the early 1970s. In 1981, NASA was the first to launch a 10-year debris program. In 1984, the *U.S. Commercial Space Launch Act* was passed. The Act clearly states that in commercial space activities, space resources should be used peacefully and the space environment protected. Although the issue of space debris is not included in the Act, the significance of the

Act is that the United States government has established a sense of space environment protection and attached great importance to this issue. In 1988, the space policy issued by President Reagan addressed the space debris. In 1989, the United States published the *Report on Orbital Debris*, 10 years earlier than the *Technical Report on Space Debris* issued by UNCOPUOS.

In 2001, the United States developed the *Government Standard Approach to Space Debris Mitigation*, which was very similar to the *IADC Mitigation Guidelines*, except that it was more explicit in its content and less stringent in its enforcement. In 2006, the Bush administration issued a new *National Space Policy* aimed at regulating the space activities. The Policy recognized the seriousness of the damage caused by orbital debris to the environment of LEO and GEO, as well as the threat of orbital debris to the safety of space objects and persons, and advocated the adoption of active and effective preventive measures to reduce the amount of space debris for the benefit of future generations. The policy is not a normative legal document, but can serve as a guide for future legislation as it regulates the space activities and emphasizes the issue of space debris.

In 2010, the US government released its new *National Space Policy*, which proposed to seek the maximum international cooperation in such areas as space environment data sharing and space debris monitoring, and to promote international space cooperation in the area of space debris monitoring on the basis of "peaceful cooperation". The policy does not have the effect of a legal document, but lays emphasis on space debris monitoring and lays a foundation for the legislation on space debris mitigation. In June 2017, an amendment to Chapter 51 of the *United States Code*, known as the *American Space Commerce Free Enterprise Act*, was introduced to Congress for debate and was passed by the House of Representatives in April 2018. This is the first time that a "space debris mitigation" program has been formally incorporated into legislation at the U.S. congressional level. The main content of the Act is to regulate private commercial launches, stipulate the approval criteria, time limit and transfer of space launch licenses, and require the establishment of an advisory committee on private outer space activities to specifically manage those activities and provide state protection for them. According to the Act, the license application must include a space debris mitigation plan as a precondition for license issuance, and the expiry date of a license is subject to the cessation of all operations approved under the license, including the implementation of the space debris mitigation plan. In the Act, a commitment to implementing the space debris mitigation plan is taken as a prerequisite for issuing a launch license so as to control the generation of space debris from the source. If further implemented by subsequent federal laws, this provision will play a significant role and have a far-reaching impact on the prevention and control of space debris pollution.

In January 2018, the US House of Representatives submitted an amendment to the *Space Debris Active Removal Act*, requiring the amendment of Section 31501 of Chapter 51 of the *United States Code* to include the active removal of space debris. The amended Section 31501 reads as follows: "The Administrator, in conjunction with the heads of other federal agencies, shall take steps to develop or acquire the technologies and policies that will enable the administration to reduce the risks associated with orbital debris and remove such debris." This is the first time the concept of space debris removal has been required to be added to a congressional bill on the books. According to the proposed amendment, the policy directive will be issued by chief executive, and the active removal, prevention and mitigation of space debris will be placed under the federal administrative agency. However, the specific institution setting needs to be determined by further amendments.

### 9.2.2 Construction of Debris-Related Laws and Regulations in Russia

As the first country engaged in outer space activities in the world, the Soviet Union made extraordinary achievements in the field of aerospace. Russia, as its successor, has gradually improved the space legislation established by the Soviet Union into a relatively mature space legislation system. The Soviet Union began to pay attention to the problem of outer space environment pollution in 1985 and drafted the first evaluation report on space environment pollution in 1990. In 1992, Russia became one of the first countries to establish a standard practical data system for the protection of space orbital resources. In 1993, Russia enacted its first law on space activities, the *Law on Space Activities of the Russian Federation* (Law on Space Activities for short), which has since undergone five major revisions. Taking the Law on Space Activities as its mother law, Russia has since adopted a series of legally binding documents, including the *Presidential Decree on the Agency for the Management of Space Activities of the Russian Federation* (1992), the *Declaration of Russian Federation Parliament on the Priority Status of the Space Policy of Russian Federation* (1993), the *Decision of the Government of Russian Federation on State Support and Guarantees for Space Activities of the Russian Federation* (1993), the *Decision of the Government of Russian Federation on the Licensing of Specific Activity Programs* (1994), the *Regulations on the Issuance of Permits for Space Activities* (1996), the *Presidential Decree on the Structure of Federal Executive Authority* (1999), and the *Law on the Conduct of Crew of the Space Station* (2002). According to statistics, Russia has issued 3 presidential decrees, 13 governmental decrees, and more than 200 departmental rules and regulations in aerospace field. Therefore, some scholars believe that Russia's space legislation activities are second to none in the world.

In addition to cataloging and continuously tracking and monitoring the space debris, Russia has successively promulgated since 2000 the *Space Debris Mitigation Standards*, the *Space Technology Program: General Requirements for Limiting the Pollution Caused by Global Space Technologies*, the *Space Technology Program: General Requirements for the Protection of Space Resources against Particle Impacts from Natural and Technological Sources* and other regulation documents to regulate the space debris mitigation.

## 9.3 Current Situation of Space Debris Legislation in China

### *9.3.1 Foundation and Progress of Space Debris Legislation*

In the 1950s and 1960s, China's outer space activities took the first step. The successful launch of the "Dong Fang Hong 1" satellite in 1970 was a landmark event indicating that China became a space state. After nearly 50 years of development, our country has made great progress in the space field and has ranked among the top of the world. Our country only started to set foot in the field of space debris in 1989. In June 1995, China joined the Inter-Agency Space Debris Coordination Committee (IADC) in the name of the China National Space Administration. However, due to the lack of resources, our space debris work at that time was limited to organizing relevant institutes to participate in international joint observation of space debris, developing a mitigation program for the third-stage of "Long March 4" rocket, and exploring the space debris databases and environmental models. In November 1999, the basic research on space debris was listed as a special national financial program thanks to the efforts of various parties. On this basis, China officially launched the *Space Debris Action Plan during the Tenth Five-Year Plan Period* at the end of 2000, which marked the entry of China's space debris research into a fast track. Since then, China has made remarkable progress in the basic research, condition construction, engineering application, international cooperation and team building of space debris domain. Especially, fruitful results have been achieved in the areas of early warning and prediction, protective design, mitigation measures and infrastructure construction. All these have laid an essential foundation for the construction of Chinese laws and regulations on space debris mitigation.

China attaches importance to the regulation of outer space activities. In terms of international law, China has always supported the UNCOPUOS and its conferences, and has taken an active part in the formulation of outer space law. At the UN Ruling Conference in 2008, China and Russia jointly submitted a draft treaty on the prevention of the weaponization of outer space. China joined the UNCOPUOS in 1980 as a member State, and joined the IADC in 1995. So far, the international space treaties to which China is a party include the *Outer Space Treaty*, the *Rescue Agreement*, the *Liability Convention* and the *Registration Convention*. In terms of regional cooperation in the field of outer space, China has also made substantial achievements. The Asia-Pacific Space Cooperation Organization (APSCO), led by China, signed the *Asia-Pacific Space Cooperation Convention* in 2005 and now has 11 member countries including Russia, South Korea and Australia. In 2015, China and other member states jointly issued the *Beijing Declaration of High-Level Forum on APSCO Development Strategy* and launched a number of technical cooperation projects. In 2018, the "High-Level Forum at the 10th Anniversary of APSCO Establishment" was held in Beijing. At the forum, the "Vision for APSCO Development in 2030" (hereinafter referred to as the Vision) was released. According to the Vision, China will adhere to the cooperation policy of "resource sharing, equality and win-win", leverage the advantages of member states, and carry out a large scale of deep-level international cooperation to provide impetus for sustainable development.

In terms of domestic laws, the largest number of outer space activities were formulated under the leadership of State Administration of Science, Technology and Industry for National Defense (the former State Commission of Science, Technology and Industry for National Defense). In 2003, the China National Space Administration (CNSA) issued the *Development Outline of the Space Debris Action Plan for 2006–2020* at the "Conference on Space Debris Research". This policy document proposes to do a good job in the protection engineering, space environmental protection engineering and detection and warning engineering for space debris, which shows that China, as a responsible power, has a positive attitude towards protecting the outer space environment and assuming the obligation of space debris mitigation. In November 2006, the research group engaged in "Legislation on Space Debris and Space Environment Protection" was established by the Legal Affairs Office of Shanghai Aerospace Administration (Shanghai Academy of Aerospace Technology) under the financial support of the former State Commission of Science, Technology and Industry for National Defense (COSTIND) to first initiate the systematic research on the legal issues of space debris in China. In this period, the Shanghai Aerospace Administration signed research contracts with Harbin Institute of Technology, Beijing University of Aeronautics and Astronautics, Beijing Institute of Technology, Shenzhen University and the Law School of Fudan University respectively, establishing cooperation channels between the space industry and the space law community. Since then, the studies on space debris have been continuously increased in the space law community. A large number of research results have emerged to effectively fill the long-time gap in China's space debris legislation research. Under the leadership of the State Administration of Science, Technology and Industry for National Defense (AOSTIND), the research group drafted a legislative proposal on space debris mitigation based on the *Measures for the Administration of Registration of Space Objects* and the *Interim Measures for the Administration of Licenses for Civil Space Launch Programs*. The proposal was specially discussed at the "National Symposium on Space debris", and was revised for six times according to the widely solicited opinions to finally form a legislative draft. In November 2007 and March 2008, the draft was discussed within the Space Debris Research Special Office and the Space Debris Research Expert Group. In May 2008 and June 2008, the consultation meetings were held in Shanghai and Beijing to solicit the opinions of space industry institutes on the draft. In February, June and September 2009, the draft was reported to the Ministry of Foreign Affairs and other relevant ministries. In November 2009, it was approved by the AOSTIND. On December 1, 2009, it was officially promulgated by the AOSTIND as the *Interim Measures for Space Debris Mitigation and Protection Management* (hereinafter referred to as the Interim Measures for Mitigation and Protection) .

In 2007, the former COSTIND officially issued China's first overall development plan for space field, namely the *11th Five-Year Plan for Space Development* (hereinafter referred to as the Plan), in which nine major scientific and technological projects were deployed. The third part of the Plan, titled "Major Tasks", proposed the implementation of "space debris warning, spacecraft protection and space debris mitigation". In addition, in order to systematically elaborate on China's outer space policy and plan the future work, the State Council has issued a white paper titled "China's Space" once every five years since 2000 to provide policy guidance for outer space activities. The *China's Space in 2016* released in late 2016 was China's fourth white paper on outer space activities. The white paper proposed five main tasks for China's future space activities, including: promoting sustainable space development; strengthening international information sharing in the new era and carrying out international cooperation in various space technology fields; refining the supporting policies for China's commercial outer space activities in the principles of "encouraging" and "orderly", and requiring to accelerate the construction of a space legislation system with space law as the core.

However, in terms of the relevant legislation, China has been lagging behind for a long time. So far, there is still no real basic law on outer space activities. In 2001, the former COSTIND and the Ministry of Foreign Affairs jointly promulgated the *Measures for the Administration of Registration of Space Objects* (hereinafter referred to as the Measures for Registration Administration), which was the first departmental regulation governing outer space activities in China and was China's space object registration system. In 2002, the former COSTIND successively promulgated the *Interim Measures for the Administration of Licenses for Civil Space Launch Programs* (hereinafter referred to as the Interim Measures for Licenses) and the supporting *Procedures for the Examination and Approval of Civil Space Launch Licenses*, which constituted the licensing system for the space behaviors of China's non-military spacecrafts. In 2010, the AOSTIND issued the *Interim Measures for Mitigation and Protection*, in which the access mechanism, prevention mechanism and management mechanism for space debris mitigation were initially established. Before that, the former COSTIND issued the technical document "Space Debris Mitigation Standard" in 2005, which stipulated the various life stages of space debris from the perspective of technical standard.

### *9.3.2 Construction of Debris-Related Laws and Regulations in China*

In 2001, the former COSTIND and the Ministry of Foreign Affairs jointly issued the *Measures for Registration Administration*, which incorporated the Registration Convention, basically corresponded to and supplemented the Convention. The Measures specified the items and types of space object registration, designated the former COSTIND as the competent department, and set the time limit for registration. The Measures prescribed the compulsory registration obligation for the space objects launched inside China and those launched outside China (for which China was a co-launching country). According to the Measures, the CNSA should, within 60 days of domestic registration, register those space objects at the UN Secretariat through the Ministry of Foreign Affairs. As a complement to the *Registration Convention* in which the time limit for registration is not set, this article can effectively urge the registration subject to actively fulfill the registration obligation. The registration of space objects implies the clarification of jurisdiction over the objects and lays a foundation for determining the attribution of debris pollution liability.

The issue of space debris mitigation was first explicitly addressed in the *Interim Measures for Licenses* released in 2002. In the Article 2 of the Measures, the COSTIND was designated as the authority in charge of civil launch activities to examine, approve and supervise the space objects to be launched. The Article 6 prescribed the launch permit mechanism, requiring the participants in outer space activities to submit applications to the competent authorities prior to the implementation of outer space programs. The impact of the pollution caused by space objects and of the issue of space debris would be reviewed to prevent the generation of space debris from the launch stage and provide institutional guarantees for the protection of outer space environment and the mitigation of space debris pollution.

The *Development Outline of the Space Debris Action Plan for 2006–2020* (hereinafter referred to as the Outline) released in 2003 is a manifestation of China's commitment to assuming the responsibility of a major power and actively preventing and controlling the space debris pollution. The Outline requires the concrete implementation of "debris protection engineering, space environmental protection engineering, and debris detection and warning engineering". Different from the White Paper, the Outline provides direct and specific regulation on the prevention and control of space debris pollution. Although the Outline mainly provides the policy-based guidance and lacks operability, it is an important step towards the legislation for space debris mitigation.

The *Space Debris Mitigation Standard* issued in 2005 explicitly required the "mitigation and control of space debris during the design, operation and after-mission disposal of space systems". The standard is widely used in spacecraft protection design, debris impact risk assessment, spacecraft avoidance route design and other areas, and plays a direct positive role in the prevention and control of space debris pollution.

The *11th Five-Year Plan for Space Development* issued in 2007 planned the "space debris warning, spacecraft protection and space debris mitigation" as main tasks, and included the debris warning engineering in the space task planning for the first time. This suggests that the issue of space debris has aroused China's attention and vigilance.

The *Interim Measures for Mitigation and Protection* issued in 2010 constitute China's first normative legal document specifically stipulating the space debris mitigation, and are of great significance to the construction of Chinese legal system for space debris pollution. The Interim Measures have initially established the spacecraft support mechanism, access mechanism, prevention mechanism and governance mechanism aimed at space debris mitigation. In terms of the support mechanism, the Article 2 makes clear that the AOSTIND is the main body to "supervise and manage the mitigation and protection of space debris generated by civil spacecrafts and vehicles, and organize the special scientific research on space debris". The Article 10 points out that "the special research programs on space debris shall be supported by the special financial funds of the State, and the State Administration of Science, Technology and Industry for National Defense shall be responsible for organizing and implementing the programs", thus providing financial support for the mitigation and control of space debris. In terms of the access mechanism, the Article 8 of the Interim Measures stipulates that the development organizations of spacecrafts and launch vehicles shall, when applying for launch licenses, put forward the "measures for space debris mitigation and protection" in the application documents, so the debris mitigation requirement becomes a mandatory threshold for spacecraft launch access. In terms of the prevention mechanism, the Article 6 of the Interim Measures focuses on the management means, requiring the spacecraft development organizations to observe the debris mitigation requirement in the whole process of engineering verification and R&D control. The Article 7 focuses on technical standard, putting forward the requirements for space debris mitigation in the process of spacecraft development and operation. In terms of the governance mechanism, the Article 13 of the Interim Measures prescribes the means of early warning and emergency handling against space debris, requiring relevant departments to carry out timely coordination and take emergency measures in accordance with the detected debris warning information.

In 2011, the Beijing University of Aeronautics and Astronautics held the "2th International Conference on Sustainable Space Development", during which the active removal of space debris was discussed as a topic. However, the active removal of space debris has not been mentioned in the normative legal documents and policy documents. The white paper "China's Space in 2016" praised China's effort in the construction of a "system of monitoring, warning, mitigation and protection technologies" for space debris, and took it as an important direction for future development. It is foreseeable that with the improvement of the basic system and mitigation system, the active removal of space debris will be included in China's agenda of debris pollution prevention and control.

At the end of 2016, the Ministry of Foreign Affairs issued a white paper titled "China's Space in 2016", which listed important progresses made in China's space sector since 2011, including "gradual improvement of the system of debris monitoring, warning, mitigation and protection technologies, and continuous improvement of the system of standards and regulations". The white paper announced that the mechanism of debris monitoring and warning had been operational. In terms of future planning, the white paper defined the construction of the debris monitoring and warning system as the main means and important goal in the protection of outer space environment. Looking ahead to the cooperation in the next five years, the monitoring, early warning, mitigation and protection of space debris was listed in the white paper as a key cooperation project. Therefore, the mitigation and control of space debris has become an important part of China's space industry, and has received full attention as the most important means of space environment protection. According to the white paper, China will further regulate the mitigation of space debris in the field of outer-space environment protection.

After a series of previous legislative efforts, the legal system for space debris mitigation and control has taken initial shape in China. China has learned from Russian centralization model in the setting of functions and powers, and achieved the connection between legislative technology standards and legal documents. This reflects China's resolution to control space debris. However, as the basic law on outer space activities is still vacant, the current regulations will face more challenges with the development of space technologies while playing a greater role [1].

## 9.4 Gaps and Development of the Construction of Space Debris Regulations

### *9.4.1 Gaps in the Construction of International Debris-Related Regulations*

Most of the five international treaties formulated by UNCOPUOS in the 20th century lack substantive provisions and provide only general principles. From this point of view, the existing international space legislation is far behind the rapid pace of the space activities carried out by various countries. Due to the limitations of historical background and interest requirements, the rules on outer space environment protection in these treaties are only the general principles, and the relevant guidelines are also characterized by voluntariness and advocacy. With regard to the protection of outer space environment, there is still no effective legal system due to the existence of legislative gaps in many areas.

1. **Lagging-behind and Insufficient Validity of the Existing Treaties**

The main legal sources of outer space environment protection are the five international treaties formulated by UNCOPUOS, namely: Outer Space Treaty, Registration Convention, Liability Convention, Moon Agreement, and Rescue Agreement. However, only a few scattered articles of the five conventions involve the protection of outer space environment. In addition, with the continuous development of space activities, the field of space activities is gradually expanded, the tasks of space activities are more far-reaching, the contents of space activities are more diverse, and the number of space activities is increasing. In comparison, the space legislation is extremely incompetent. In the face of the new situation in space activities, the existing space legislation shows not only a delayed response but also a lack of operability. The *Outer Space Treaty* is the fruit of ten years of human's space activities. As the first legally-effective component of the Outer Space Law, its fundamental status and landmark significance are self-evident. Although the Treaty clearly points out the principle of environmental protection and requires all countries to avoid causing damage to the environment in outer space activities, the rules in the Treaty are mostly the principles rather than effective measures for solving the problem of space debris. The *Liability Convention* defines the subject of liability, subject of claim, and rules and principles for environmental damages caused by space objects, which, however, are only applicable to the damages within the sovereignty scope of sovereign states and are not applicable to the damages caused by space objects to the outer space environment[2].

In addition, the *IADC Mitigation Guidelines* and the *UNCOPUOS Mitigation Guidelines*, as the international documents of advocating or programmatic nature, prescribe only the purposes or basic principles of environmental protection and seldom address the implementation of laws. They do contribute to the establishment of international treaties and customs and important principles of international environmental law, exert certain influence on the behaviors of international law subjects, play an important role in environmental protection, and effectively promote the development of international environmental law. However, they are not legally binding and enforceable, so their effectiveness will be greatly reduced.

2. **Vagueness of Related Concepts Such as "Space Debris"**

When it comes to related concepts, "space debris" should be the term most needful of a clear definition. However, most of the international treaties listed above use the term "space object", but do not mention the term "space debris" or clarify its connotation. There are few legal provisions for regulating space debris for the following main reasons. First, the above international conventions were mostly signed in the middle and late 20th century. During that period, there were not many countries capable of space launch. Even the two major space powers, the United States and the Soviet Union, were limited by many factors such as science & technology and funds, so the space launch activities were not very frequent. The frequency of collisions between space objects was not very high, and the occupation of space resources by space objects was not very serious. Therefore, space debris was not produced too much, and its hazard easily escaped people's notice. The international treaties signed at that time failed to take full account of the legal regulations on space debris, which was also a manifestation of their legal lag. Second, these treaties were formulated against a special historical background. In the middle and late 20th century, the United States and the Soviet Union fought for hegemony and created a confrontation situation with fierce competition in political, economic and military aspects. As the major participants in space activities, they were too busy to make international conventions, so they could not make substantive contributions in this aspect.

The ambiguity is not solely reflected in the term "space debris". In the *Outer Space Treaty*, the applications of the terms such as "harmful pollution", "adverse changes to the environment" and "potential harmful interference" have not been clearly explained, which will inevitably leave hidden dangers for the implementation of the Treaty [2]. According to Article 1 of the Treaty, all States Parties have the right to freely develop and utilize outer space resources, and outer space is the common property of mankind. There is nothing wrong for the article to grant all States Parties the freedom to develop and utilize space resources, but the article fails to impose appropriate and necessary restrictions on such freedom. This defect is bound to lead to the tragedy of the Commons. All space countries blindly pursue the "free" development and utilization of outer space resources while ignoring the protection of outer space environment, resulting in the formation of an outer space being flooded with a large amount of space debris. This has not only seriously damaged the outer space environment, but also greatly hindered human's development and utilization of outer space.

3. **Difficult Identification of the Subject of Liability and Lack of the Subject of Claim**

The outer space is too vast for the existing technologies to comprehensively monitor the entire outer space region with human activities. Even relying on the current *Registration Convention*, the true owners of some space objects (especially space debris) are still difficult to identify. Meanwhile, the damages caused by such objects are increasing day by day. In this case, how to protect the outer space environment and how to identify the subject of liability is still a legal blank. To regulate this issue legally, the key is to identify the subject of liability. This is important for solving the problem of environmental damage caused by space debris [2].

According to the existing international space law, the launching State is the only subject liable for the damage caused by space debris. In accordance with the *Registration Convention*, the launching State is required to enter the outer space object into its register. Where there are more than two launching States, they shall, through negotiation, choose one of them for registration. The chosen State shall be the registered State and exercise jurisdiction and control over the space object launched by them into outer space. According to the *Registration Convention*, where there are more than two launching States, they shall, through negotiation, choose one of them for registration, without prejudice to the provisions in the agreements reached among them concerning the jurisdiction and control over the outer space object launched by them and the persons onboard. Therefore, there may be a complex situation in which the ownership of an orbiting satellite changes, but the State of registry recorded in the register does not change in time, and the State with real control of the satellite does not change into the subject liable for debris-caused damages.

With the development of science and technology, the outer space activities carried out by various countries have been commercialized and privatized. More and more private entities and international organizations are engaged in satellite launching activities. It is unrealistic to require the States to control the space operation of many private entities. So far, there is no clear regulation on who should be held liable when a private entity causes damage. Therefore, in order to adapt to the new situation, the international liability for environmental damages caused by private entities should be included in the agenda of space legislation. The claims for cross-boundary environmental damages cover the damage to the environment within the jurisdiction of a State or caused by the misconduct of another State, and the damage to the environment outside the jurisdiction of a State. The case that the space debris causes material damage to the environment of another State undoubtedly falls into the first damage category, where the victimized State can be regarded as the subject of claim. However, the serious damage caused by space debris to the outer space environment belongs to the second category. How to determine the subject of claim in this case or which subject should represent human beings as the subject of claim is not stipulated in the current space law. Therefore, the lack of the subject of claim for any damage to the commons not under the jurisdiction of any country is a major problem that needs to be solved urgently [3].

### 9.4.2 Improvement and Development of Chinese Laws and Regulations on Space Debris

After many years of construction, the legal regime for space debris mitigation has been initially established in China, reflecting the responsible image of our country as a responsible great power. However, due to the limitations of space technology level and space law research ability, China's legal regime for space debris mitigation is still extensive and needs to be greatly improved.

1. **Adhere to Basic Principles**

It should be recognized that the foundation of space debris research in China is still weak. Our capability of autonomous debris observation is still difficult to meet the requirements of space debris warning. There is still a big gap in the development of advanced protective materials and the construction of ground test equipment. In addition, although China's space industry has made great progress, it is still in the incubation and development period and needs a relatively relaxed policy environment. Therefore, the aggressive space debris mitigation without considering our national situation is likely to produce a negative impact on the overall development of China's space industry. This is an important factor we must consider when constructing the legal regime for space debris mitigation and control. At the same time, we should not negate or exclude our international responsibilities while considering only our economic interests. We should be fully aware of the far-reaching impact of space debris issue, and fully recognize the necessity and urgency of space debris mitigation and control to ensure the long-term interests of our space industry [4].

2. **Grasp Key Directions**

The legal regime for space debris pollution should cover at least five aspects including: rights and obligations organization, constraint permission, early warning/prevention/control, resource guarantee, and responsibility investigation. Among them, the organization of rights and obligations has been made clear in the *Interim Measures for Mitigation and Protection*. The constraint permission is outlined in the *Interim Measures for Mitigation and Protection*, but is detailed in the *Interim Measures for Licenses*. The technical standards supporting the early warning/prevention/control have been detailed in the *Interim Measures for Mitigation and Protection*. The principles of resource support have been given in the *Interim Measures for Mitigation and Protection*. Such flexible provisions are considered acceptable as resource support is related to finance. However, the accountability mechanism for space debris pollution is completely absent from the existing regulations, which obviously cannot meet the practical need of debris pollution prevention and control. In the future, the construction of accountability mechanism will become a key development direction of China's legal regulations on space debris pollution [1].

## References

1. Sun, X. (2019). *Comments on China's legal regime for space debris pollution.* Zhongnan University of Economics and Law.
2. Hou, W. (2014). *Study on international liability for environmental damages caused by space debris.* Beijing Jiaotong University.
3. Guan, L. (2016). *International liability for environmental damages caused by space debris.* Liaoning University.
4. Feng, G. (2013). *Research on the legal mechanism for space debris mitigation.* East China University of Political Science and Law.

# Index





J. Zhang et al., *Space Debris Research*, https://doi.org/10.1007/978-981-96-9570-6

## T